GEOLOGY UNDERFOOT IN SOUTHERN CALIFORNIA

Second Edition

Arthur Gibbs Sylvester
Robert P. Sharp
Allen F. Glazner

ILLUSTRATED BY
Elizabeth O'Black Gans

2020
Mountain Press Publishing Company
Missoula, Montana

The Geology Underfoot series presents geology with a hands-on, get-out-of-the-car approach. A formal background in geology is not required for enjoyment.

First Printing, April 2020
Second Printing, August 2024

ON FRONT COVER
Red Rock Canyon amphitheater, Red Rock Canyon State Park
(from photograph by Arthur G. Sylvester)

Cover design by Jeannie Painter

All photographs by the authors unless otherwise credited.

Library of Congress Cataloging-in-Publication Data

Names: Sylvester, Arthur G., author. | Glazner, Allen F., author. | Sharp, Robert P. (Robert Phillip), author.
Title: Geology underfoot in southern California / Arthur Gibbs Sylvester, Robert P. Sharp, Allen F. Glazner ; illustrated by Elizabeth O'Black Gans.
Description: Second edition. | Missoula, Montana : Mountain Press Publishing Company, 2020. | Series: Geology underfoot | Includes bibliographical references and index. | Summary: "First published in 1993, this groundbreaking book is now fully revised and updated with the latest scientific information. Join a team of geologists as they use clear prose, concise illustrations, and dramatic full-color photographs to tell the stories of 21 amazing geologic sites, 3 of which are completely new to the book-San Andreas Fault, Devils Punch Bowl, and St. Francis Dam. Reading the rocks like pages in a book, Geology Underfoot in Southern California offers an inside view of the southland's active and sometimes enigmatic landscape"—Provided by publisher.
Identifiers: LCCN 2020005667 | ISBN 9780878426980 (paperback)
Subjects: LCSH: Geology—California, Southern.
Classification: LCC QE90.S65 S53 2020 | DDC 557.94/9—dc23
LC record available at https://lccn.loc.gov/2020005667

Printed in the United States

P.O. Box 2399 • Missoula, MT 59806 • 406-728-1900
800-234-5308 • info@mtnpress.com
www.mountain-press.com

If a stone appeals to me and elevates me, tells me how many miles I have come, how many miles remain to travel, and more the better, if it reveals the future to me in some measure, it is a matter of private rejoicing. If it did the same service to all, it might well be a matter of public rejoicing.

—HENRY DAVID THOREAU, 1856, IN *THE JOURNAL*

Rocks, like everything else, are subject to change . . . and so also are our ideas about them.

—F. Y. LOEWINSON-LESSING, 1936

The sublunary world is divided for the alchemist into three kingdoms: the mineral kingdom, the vegetable kingdom, and the animal kingdom. . . . The rhythm of the animal kingdom is that of everyday existence. The rhythm of the mineral kingdom is that of the ages, of life calculated in millennia. As soon as we contemplate the thousands of years of existence for minerals, cosmic dreams come to us.

—GILBERT WHITE, 1773, IN *THE NATURAL HISTORY AND ANTIQUITIES OF SELBOURNE*

Even before I accepted what I saw, I heard the silence, felt it, like something solid, face to face . . . . A silence so profound that the whole colossal chaos of rock and space and color seemed to have sunk beneath it and to lie there cut off, timeless . . . . I knew that something had happened to the way I looked at things.

—COLIN FLETCHER, 1967, IN *THE MAN WHO WALKED THROUGH TIME*

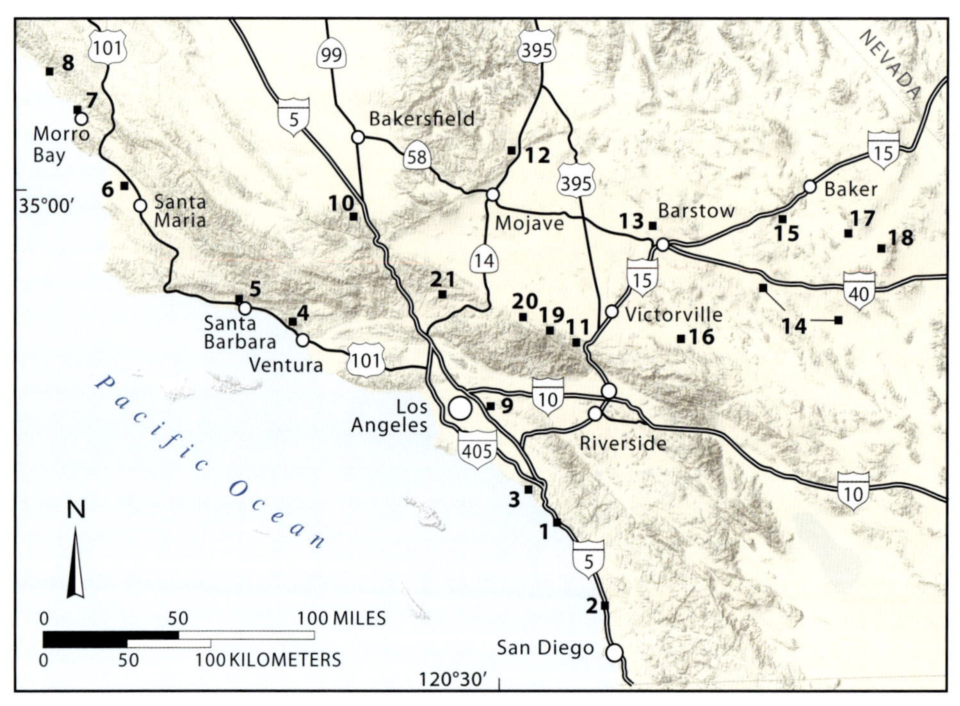

Location of vignettes in Geology Underfoot in Southern California.

Contents

Preface to the First Edition

A vignette can be an ornamental sketch, a striking picture, a vivid scene, or a short story. Vignettes presented herein are stories, like snapshots, focused on some particular scene, relationship, or feature selected from the rich mixture of southern California's geological phenomena. Like snapshots, they do not cover everything, only items in focus and within the field of view.

Each vignette treats a geological subject of particular interest and significance. Some involve features well-known to many readers. Others treat relationships with which readers may be unacquainted or to which they have not given previous thought. Some treat matters of practical concern: San Onofre nuclear plant, Santa Barbara Harbor, and Ventura Avenue oil field. Many offer pure education and enjoyment: Rainbow Basin, Torrey Pines, San Simeon beach pebbles. A few involve some limited, largely easy walking; some require careful observation; some can be appreciated from the car.

Topics selected are geographically dispersed but purposely located near heavily populated or frequently visited areas. Some extensively documented areas, such as Joshua Tree and Death Valley National Parks, are not treated. The spectrum of available geological vignettes is so broad and varied that some readers are bound to be disappointed their favorite spot is not included here. To them we apologize. Selecting just twenty-one out of a hundred or more outstanding candidates was a traumatic task. Our aim is to present a description and discussion of geologic features in simple conversational style, comfortable for nonprofessional readers. We have tried to keep technical jargon to a minimum. It is difficult to communicate about geologic phenomena, however, without some rudimentary knowledge of a few basic principles. Such information follows. Reference to the comprehensive glossary also helps. For each vignette we show the local geography with a map or two. It is helpful to also bring along good local road maps, such as those published by the Automobile Club of Southern California, US Geological Survey topographic maps, which are available online, and images from Google Earth.

Material for vignettes has been drawn from many publications and field trips, too numerous to acknowledge individually. The authors of all are saluted with warm thanks. We enjoyed putting this volume together. It is designed to give readers pleasure in becoming acquainted with natural features of the world around us.

—Robert P. Sharp and Allen F. Glazner

Preface to the Second Edition

It has been twenty-five years since publication of the first edition of this book, which inaugurated the Geology Underfoot series at Mountain Press. We have endeavored to include in this Second Edition some of the enormous amount of new information that has become available about southern California geology in the intervening years without damaging the book's intent and style.

Sylvester revisited all of the sites to verify or update site access altered or precluded by increased urbanization and revised administrative policies and to obtain color images to replace black and white photos of the First Edition.

We are indebted to the very many geologists on whose original research we have depended so heavily in researching and writing this book. We benefited greatly and learned much from discussions with, and manuscript reviews by, Bill Elliott, Jeff Miller, Jim Boles, Nancy Riggs, Alex Simms, Nate Onderdonk, Doug Burbank, Fred and Judy Chester, Brad Hacker, Ray Weldon, Bruce Lander, Kathryn Bowers, Steve Lipshie, Craig Nicholson, Tom Heaton, and Elaine Hanford. Many heartfelt thanks to all of them.

Although the authors tend to get most of the credit, any highly illustrated book like this one is in reality a team effort. Therefore, it is a real pleasure to acknowledge and thank Libby Gans for her masterful work in preparing the maps and illustrations, and we're deeply grateful to all of the Mountain Press production team who have worked to make this book a success, especially Jenn Carey for editorial guidance, and designer Jeannie Painter. Sylvester acknowledges the support of a UCSB Dickson Emeritus Fellowship.

As we write this, the *Curiosity* rover/robot-geologist continues its climb up the slopes of Mount Sharp (Aeolis Mons) on Mars. Most of all we thank our coauthor, the late Bob Sharp, after whom that mountain was named, for showing us how to engage the public in geologic thought. We look forward to a *Geology Underfoot on Mars* series.

—Arthur G. Sylvester and Allen F. Glazner, 2020

An Important Note to Readers

Although the publisher and authors have taken reasonable steps to ensure the accuracy and timeliness of the information contained in the book, readers are strongly encouraged to confirm details before making any travel plans. Location and direction information may change; GPS coordinates are approximations and should be treated as such. If you discover any out-of-date or incorrect information in the book, please let us know by sending an email to the publisher at info@mtnpress.com.

Geology Underfoot in Southern California was written in the spirit of adventure, and readers are cautioned to travel at their own risk and obey all local laws. Some of the places described in this book are not open to the public and are not meant to be visited without appropriate permissions. Neither the authors nor the publisher shall be liable or responsible for any loss, injury, or damage allegedly arising from any information or suggestions contained in this book.

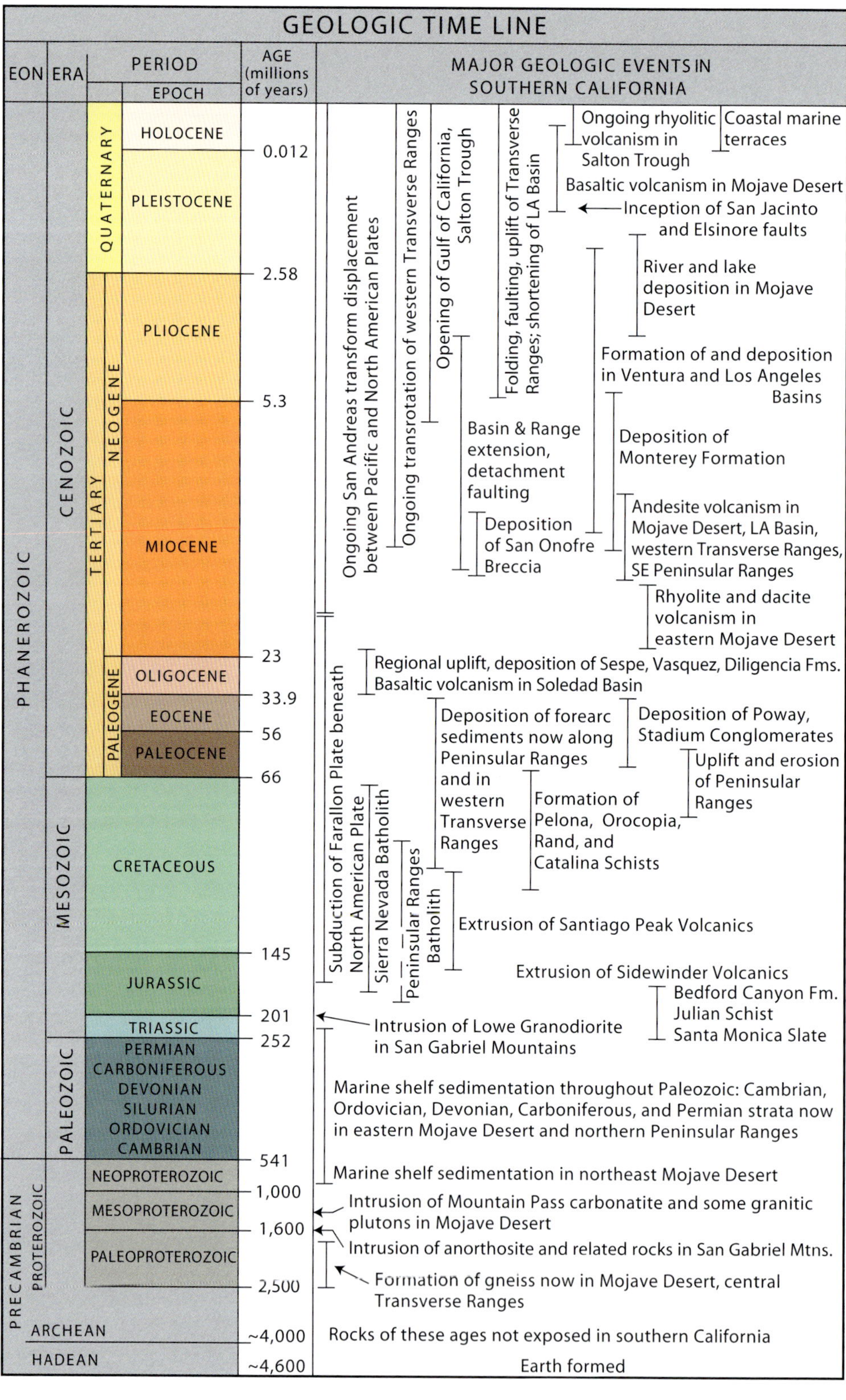
GEOLOGIC TIME LINE
EON
ERA
PERIOD
EPOCH
AGE (millions of years)
MAJOR GEOLOGIC EVENTS IN SOUTHERN CALIFORNIA
PHANEROZOIC
CENOZOIC
QUATERNARY
HOLOCENE
0.012
PLEISTOCENE
2.58
TERTIARY
NEOGENE
PLIOCENE
5.3
MIOCENE
23
PALEOGENE
OLIGOCENE
33.9
EOCENE
56
PALEOCENE
66
MESOZOIC
CRETACEOUS
145
JURASSIC
201
TRIASSIC
252
PALEOZOIC
PERMIAN
CARBONIFEROUS
DEVONIAN
SILURIAN
ORDOVICIAN
CAMBRIAN
541
PRECAMBRIAN
PROTEROZOIC
NEOPROTEROZOIC
1,000
MESOPROTEROZOIC
1,600
PALEOPROTEROZOIC
2,500
ARCHEAN
~4,000
HADEAN
~4,600
Ongoing San Andreas transform displacement between Pacific and North American Plates
Ongoing transrotation of western Transverse Ranges
Opening of Gulf of California, Salton Trough
Folding, faulting, uplift of Transverse Ranges; shortening of LA Basin
Ongoing rhyolitic volcanism in Salton Trough
Coastal marine terraces
Basaltic volcanism in Mojave Desert
Inception of San Jacinto and Elsinore faults
River and lake deposition in Mojave Desert
Formation of and deposition in Ventura and Los Angeles Basins
Basin & Range extension, detachment faulting
Deposition of Monterey Formation
Deposition of San Onofre Breccia
Andesite volcanism in Mojave Desert, LA Basin, western Transverse Ranges, SE Peninsular Ranges
Rhyolite and dacite volcanism in eastern Mojave Desert
Regional uplift, deposition of Sespe, Vasquez, Diligencia Fms.
Basaltic volcanism in Soledad Basin
Subduction of Farallon Plate beneath North American Plate
Sierra Nevada Batholith
Peninsular Ranges Batholith
Deposition of forearc sediments now along Peninsular Ranges and in western Transverse Ranges
Deposition of Poway, Stadium Conglomerates
Uplift and erosion of Peninsular Ranges
Formation of Pelona, Orocopia, Rand, and Catalina Schists
Extrusion of Santiago Peak Volcanics
Extrusion of Sidewinder Volcanics
Bedford Canyon Fm.
Julian Schist
Santa Monica Slate
Intrusion of Lowe Granodiorite in San Gabriel Mountains
Marine shelf sedimentation throughout Paleozoic: Cambrian, Ordovician, Devonian, Carboniferous, and Permian strata now in eastern Mojave Desert and northern Peninsular Ranges
Marine shelf sedimentation in northeast Mojave Desert
Intrusion of Mountain Pass carbonatite and some granitic plutons in Mojave Desert
Intrusion of anorthosite and related rocks in San Gabriel Mtns.
Formation of gneiss now in Mojave Desert, central Transverse Ranges
Rocks of these ages not exposed in southern California
Earth formed

THE BIG PICTURE

The Danish physician Petrus Severinus (1542–1602) urged his students to "get away to the mountains, the deserts, and the deepest recesses of the Earth. In this way and no other will you gain a true knowledge of things and of their properties." Severinus knew that rocks and rock structures tell geologic stories about the Earth that can be read like pages in a book. Rocks record processes that created and continue to shape our modern landscapes—the geology underfoot. And if rocks and rock structures tell the stories, then the kinds of rocks provide the paragraphs and sentences, and minerals are the words.

Rocks come in three main varieties—igneous, sedimentary, and metamorphic—and we can deduce their origins from rock textures and the kinds of minerals in them. Igneous rocks form as magma cools and solidifies. They are primary, the source of all other rocks, because the Earth was entirely molten early in its history. To quote an old professor, who each year began his introductory geology course with this solemn announcement, "In the beginning, all things were igneous."

Some igneous rocks crystallize below the surface; others do so at the surface after they erupt from volcanoes. Either way, they vary greatly in composition and texture. Granite and rhyolite, for example, have the same chemical composition, but they look quite different. Granite is a speckled rock with black, pink, and white crystals large enough that you can easily see them without a magnifier. It may require thousands to millions of years of cooling thousands of feet below the Earth's surface to produce crystals this big. Rhyolite is a pale tan or pink volcanic rock that erupts upon the Earth's surface and consists mostly of microscopic crystals or glass. Granite and rhyolite are both rich in silicon and poor in iron and magnesium.

Basalt, another igneous rock, is the most common kind of volcanic rock on the Earth's surface; it forms all ocean crust, the bedrock beneath unconsolidated sediments that settled to the ocean floor. Like rhyolite, basalt is a mass of microscopic crystals. Unlike rhyolite, basalt is black, and it contains much less silicon than granite and rhyolite but a great deal more iron and magnesium.

The mechanical breakup and chemical decomposition of rocks—any kind of rock—makes sediment, which can be transported by rivers, streams, glaciers, and waves. It is deposited in layers and eventually hardens into sedimentary rock. Consider sandstone. It starts out as layers of sand deposited wherever sand can accumulate, such as on a floodplain, on a beach, or in a sand dune field. If the sand is buried by more sediment and if groundwater passes through it depositing cementing minerals, the sand grains stick together and end up as sandstone.

Metamorphic rocks form through the recrystallization of other rocks, largely without melting. Heat is the primary agent; all metamorphic rocks recrystallize at fairly high temperature, many at something approaching a red heat. High heat is generally accompanied by high pressure deep in the Earth's crust. Gneiss and schist are good examples of metamorphic rocks common in many parts of southern California.

Common southern California rocks. Upper row, left to right: granite, rhyolite, and basalt. Lower row, left to right: sandstone, schist, and gneiss.

A Look Within

Rocks and the landscapes that contain them are influenced by the internal forces that deform the Earth's crust. The engine providing those internal forces derives from deep within the Earth's core, which consists of two parts. The ball-shaped inner core, made of solid, superhot iron with a diameter of roughly 760 miles, is encased within the outer core, a 1,370-mile-thick layer composed of a liquid nickel-iron alloy. The semisolid mantle, about 1,800 miles thick, encloses the outer core. Outside the mantle is the crust on which we live, no thicker in proportion to the size of the Earth than the skin on a large apple.

Mantle rocks are slightly—very slightly—radioactive. The decay of radioactive elements produces heat that slowly makes its way to the Earth's surface and is eventually radiated to space. To help shed this heat, the mantle flows very slowly, producing great convection cells wherein hot mantle rises, moves laterally as it cools, and then sinks back into the lower mantle. Even though mantle material is extremely viscous and moves only a few inches per year, that movement is fast enough to provide the forces that deform the overlying Earth's crust into mountains and ocean basins. The movement has also broken the crust into great rigid pieces called plates.

The outermost rigid part of the Earth, about 60 miles thick, consists of a dozen or so plates, depending upon who does the counting. Some of the plates are huge, others quite small. In southern California, we think mostly of the Pacific Plate, which underlies most of the Pacific Ocean, and the North American Plate, which extends from the Pacific coast to the middle of the Atlantic Ocean. Both plates include areas of continental and oceanic crust as their passengers. Oceanic crust is normally about 3 miles thick, continental crust about 20 miles thick. The rest of the plate, down to a depth of about 60 miles, is upper mantle rock. The crust and this uppermost part of the mantle are called the lithosphere. The base of the lithosphere is not well-defined but is the zone where mechanical behavior changes from brittle fracturing to plastic flow at inches per year. Rocks in that outer zone are cooler than those below, and therefore more rigid and brittle. Think of the lithosphere as an outer rind on the Earth like the sugar coating on a candy apple.

Beneath the rigid plates lies the asthenosphere, a weak layer wherein the mantle is hotter, weaker, and, in places, partially molten. This weak zone allows plates of the lithosphere to move. Different plates travel at different speeds; some poke along at a fraction of an inch per year, others speed along at as much as 4 inches per year. Two inches per year, about the rate your fingernails grow, is the average rate.

Each plate moves in its own direction, toward its own destination. Plates move like bumper cars in a complicated pattern, pulling away from each other, colliding with each other, or sliding past each other. Enormous underwater mountain ranges form where plates pull away from each other—the Mid-Atlantic Ridge, for example, and the East Pacific Rise in the eastern Pacific Ocean. Fissures open in the crest of the ridge as the plates separate, and molten basaltic lava erupts to heal those fissures and thereby make new oceanic crust.

Where plates collide, long trench-like depressions form on the ocean floor, and chains of volcanoes may erupt 50 to 100 miles away from but parallel to the trench. The Aleutian trench and its associated volcanoes are a good example.

Watching a collision between two lithospheric plates is probably as close as we will ever come to seeing the mythical encounter between an irresistible force and an immovable object. The Earth solves that dilemma by letting one plate sink beneath the other into the mantle. The plate that sinks invariably has oceanic crust on its surface. Oceanic slabs sink beneath continental ones because they are denser; when two oceanic plates meet, one sinks and the other doesn't. Oceanic trenches are where the sinking plate bends down to start its long plunge into the hot depths of the Earth. Geologists call these trenches subduction zones. They swallow old oceanic crust at the same rate that eruptions at oceanic ridges create new crust. Subduction normally causes a lot of igneous activity within the overriding plate.

Where plates slide horizontally past each other—whether deep underwater or on land—they create enormous faults, such as the San Andreas fault. Geologists call such zones of horizontal slip between plates transform plate boundaries. Although they do not spawn volcanic chains, transform boundaries do cause earthquakes as the plates on either side drag alongside each other, crumpling adjacent rock layers into folds and breaking them along faults.

The San Andreas fault slices through the southern one third of California, separating the Pacific Plate on the west from the North American Plate on the east. The Pacific Plate moves northwest relative to the North American Plate at a rate of nearly 2 inches per year. Geologists call the San Andreas a right-lateral strike-slip fault based on its sense of displacement: if you stand on one side of the fault and look across it, the opposite side moves to your right. In other words, Los Angeles, on the Pacific Plate, is approaching San Francisco, on the North American Plate, in short jerks accompanied by earthquakes. The San Andreas fault provides Californians with a dynamic and unstable geologic environment—but a stunning natural landscape.

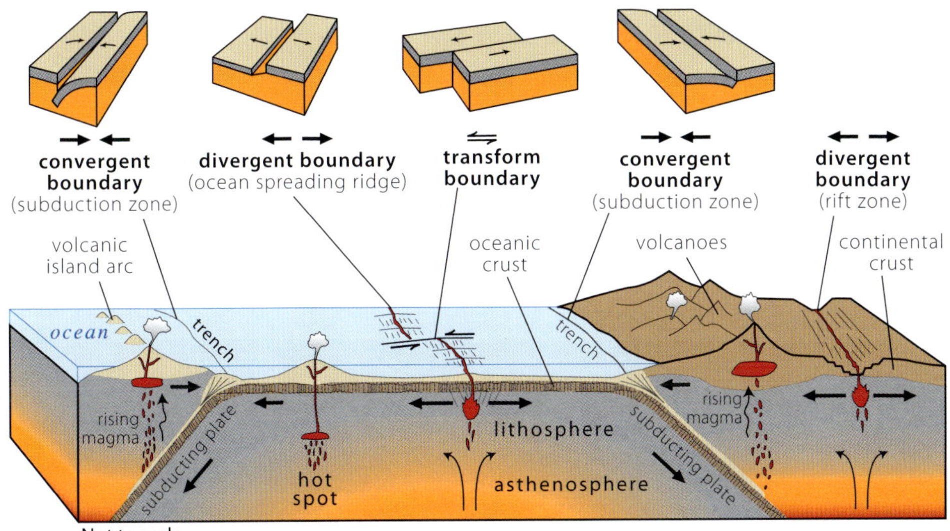

A slice of the Earth's crust and mantle shows relationships among the main structural elements of the lithosphere, wherein tectonic plates converge, diverge, and slip parallel to one another. The boundaries between converging plates are major faults.

Keeping Geologic Time

The question of time pervades every geologic discussion. The solar system and the Earth are about 4.6 billion years old, an incomprehensibly long time, even to hardened geologists who think about it regularly. Geologists have divided that yawning abyss of time into many subdivisions based on profound changes in the fossil record. From oldest to youngest, the four major divisions are: **Precambrian**, which lasted from 4.6 billion to 541 million years ago; **Paleozoic**, which lasted from 541 to 252 million years ago; **Mesozoic**, which lasted from 252 to 66 million years ago; and **Cenozoic**, from 66 million years ago to present.

To get the sense of the depth of geologic time, let's squeeze it into one human year. Nothing is preserved on Earth that tells us what happened during all of January and February. Not until July 17 does multicellular life begin to show up after a lot of rock has formed and eroded away since the end of February. Most of the year passes without life on land until early November, when land plants arrive. Then on November 18, **Precambrian** time ends and complex life bursts forth (541 million years ago). Toxic gases from Siberian volcanoes wipe out 96 percent of life on December 12 at the end of **Paleozoic** time (252 million years ago). Dinosaurs appear on December 13, but most of them go extinct on December 26 when a Manhattan-size asteroid strikes the Yucatán peninsula and gargantuan eruptions occur in India at the end of the **Mesozoic** era (66 million years ago). Grass and animals spread across the dinosaur-free land over the next five days of **Cenozoic** time. Human beings begin to put some land underfoot at 5:18 pm on December 31. Glacial ice ages come and go in the last few hours of that day. Americans sign the Declaration of Independence in the final two seconds of the year.

Geologic time is very long. Given time in geologic measure, processes that seem painfully slow to us, such as the erosion of hard rock by flowing water, can work profound changes. In these vignettes, we focus mainly on events of Cenozoic time and the rocks and the landscapes they created. However, California contains rocks from many eras.

Geologists measure geologic time in many ways, but the simplest and most reliable way is to analyze the decay of radioactive elements into other elements, a type of geologic "clock" that runs within rocks. The proportion of the radioactive parent element to the stable daughter product measures the age of the mineral or rock. To determine the age of a rock is a matter of precise chemical analysis and knowledge of decay rates determined from laboratory studies. The most useful clocks involve the slow disintegration of potassium to argon and uranium to lead.

Geologic History of Southern California

The pages of southern California's geologic history book go back more than 1.6 billion years, to a time when granite and metamorphic rocks formed at the margins of an ancient continent. Little is known about this early time because only small remnants of those old rocks remain. The well-displayed record began about 800 million years ago, when a supercontinent that contained California's oldest rocks split to create a new ocean. The piece that split away may now be part of Antarctica, or Australia, or China—those puzzle pieces have yet to be put together in a way that satisfies everyone. Sediments, which later became sedimentary rocks, were laid down along the new continental margin throughout Paleozoic time. In places, these rocks accumulated to thicknesses of several miles as the continental margin slowly subsided. Mountain ranges in the Death Valley region of eastern California contain thick layers of sedimentary rocks that date back to this time. The caves of Mitchell Caverns (Vignette 18) are also sculpted in Paleozoic rocks.

During Paleozoic time, southern California was in a situation not unlike a mirror image of the shallow continental shelf off the east coast of the United States today—drowned in shallow water and slowly taking on a greater and greater load of sediment. This quiet, depositional coastline extended from what is now the Mojave Desert through southern Nevada to northern Utah. The shoreline fluctuated greatly as sea level rose and fell, sometimes moving as far inland as what is now the Grand Canyon.

About 250 million years ago, around the beginning of Mesozoic time, the situation changed drastically. A new subducting boundary formed as plates started to dive beneath the new western edge of North America. This massive plate reorganization dramatically changed the landscape. The Farallon Plate lay west of and was subducting beneath what is now California. A north-south chain of vigorous volcanoes erupted along the length of the state east of the subduction zone. The great granite batholiths of the Sierra Nevada, Peninsular Ranges, and Mojave Desert are the roots of those old volcanoes, and remnants of the volcanoes themselves can be found in many parts of California.

Much of the crust of present-day California was built by the new volcanism, which thickened the crust and produced dry land. Rocks in the Coast Ranges and Franciscan Formation (Vignettes 3 and 8) were once sediments that were scraped off oceanic plates as they subducted

Granite typically fractures and then decomposes by chemical weathering into boulder piles like these in the Granite Mountains near Kelso Dunes. This granite is part of the great Mesozoic batholithic intrusions of the Mojave Desert and Peninsular Ranges. (34°48.513N, 115°37.187W)

beneath North America. That Mesozoic subduction and construction of the Sierra Nevada and Peninsular Ranges continued until about 75 million years ago, when volcanism stopped, perhaps because the oceanic plates plunged beneath North America at a shallower angle and didn't generate magma.

The early Cenozoic history of California, from 65 to about 25 million years ago, is rather fuzzy—few rocks of this age exist in inland southern California, so there is little record of what happened. Some marine sedimentary rocks of this age are exposed, however, along the San Diego coast (vignette 2) and in the western Transverse Ranges. By all accounts, California got a much-needed rest after the Mesozoic magmatic madness. It needed to rest up for what was to come.

A massive blast of volcanism and faulting ended the early Cenozoic quiet time about 30 to 25 million years ago. Subduction ceased along the California coast as the Farallon Plate was nearly completely consumed and the North American Plate became juxtaposed with the Pacific Plate. A shearing motion between the two plates caused the crust of what is now Nevada and eastern California to stretch and break along great faults. The volcanism and extensional faulting moved from south to north at a few inches per year, passing through southernmost California about 30 to 25 million years ago (Vignette 7) and the Mojave Desert about 16 million years ago. The faulting was a particular kind in which the crust stretched dramatically westward to make mountains and valleys, like the Panamint Mountains and Death Valley, both parts of the Basin and Range geologic province. Small Mojave basins produced by this active extension, such as Rainbow and El Paso Basins (Vignettes 12 and 13), filled with lakes and sediment in recent geologic times.

As these new forces were applied to the region, the San Andreas fault system began to develop about 25 million years ago. The south end of the San Andreas fault enters the Salton Trough, the basin at the northern end of the Gulf of California, which opened as the Pacific Plate captured and dragged Baja California away from the Mexican mainland. At the same time, the slice of coastal California west of the fault, referred to as Alta California, was also kidnapped and dragged by the Pacific Plate. The north end of the San Andreas fault, at Cape Mendocino in northern California, meets the south end of the still existing subduction zone along the Washington and Oregon coasts.

One result of all this slicing, dicing, and kidnapping is the formation of the Transverse Ranges, an imposing chain of east-west trending mountains, including the San Gabriel, Santa Monica, and Santa Ynez

Mountains and the Channel Islands. (The Santa Barbara Channel is a submerged part of the Transverse Ranges.) They are called "transverse" because their east-west trend is contrary to the general north and northwest trend of all other mountain ranges in the western United States, except for the enigmatic Uinta Mountains in northeastern Utah.

Geologists were long aware of the similarity of pre- and early Cenozoic coastal sedimentary rocks along the edge of the California Peninsular Ranges between San Diego and Los Angeles and those in the western Transverse Ranges near Santa Barbara, but they were unable to reconcile their connection. From the interpretation of stream current directions in the rocks, it eventually became clear that the coastal sediments between San Diego and Los Angeles were transported from east to west from a continental source. Those in the Santa Ynez Mountains, however, seemed to have been transported from south to north from an unknown, cryptic source.

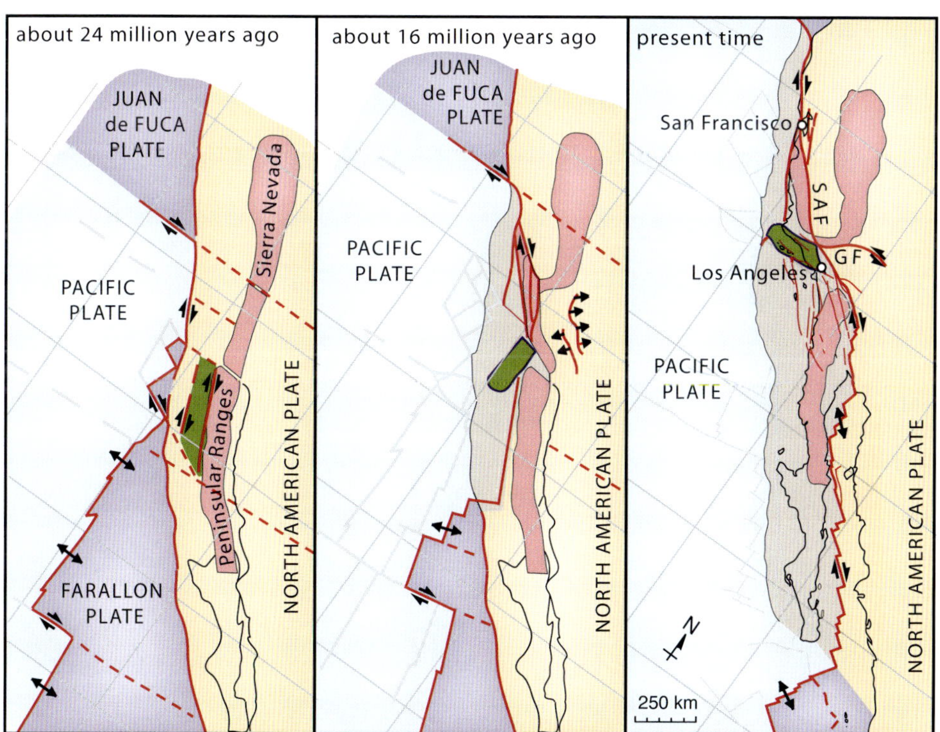

Tectonic model for Pacific and North American Plate interactions since 24 million years ago and the rotation of the western Transverse Ranges (green patch). SAF is San Andreas fault; GF is Garlock fault. Tan is area captured by the Pacific Plate. Pink is granitic batholiths. Faults are red. —After Nicholson and others, 1994

Geologists also mapped a unique suite of Eocene volcanic cobbles and boulders eroded from volcanoes in northern Mexico and transported by a major river system from there through San Diego County to form bold outcrops on the coast at La Jolla. Those same very distinctive rocks are found nowhere else in California *except* for outcrops on the Channel Islands of Santa Rosa and Santa Cruz.

Another clue came from Jurassic-age lava. Volcanic rocks contain the mineral magnetite, which orients itself toward the North Pole at the time of eruption. The magnetic direction recorded in Peninsular Ranges volcanic rocks of Jurassic age is north-south, as it is throughout almost all of California, but it is east-west in the Transverse Ranges rocks.

When these and other conflicting and contrary observations were put together, the explanation was inescapable: The Pacific Plate nipped off a piece of the North American Plate—the slab of California west of the San Andreas fault—about 16 million years ago and carried it northwestward with the Pacific Plate. Along the way, a piece in that slab, the Transverse Ranges, snagged its east end so that the remainder rotated about 100 degrees clockwise like a log caught by an eddy in a stream. Today, that kidnapped piece of California west of the San Andreas fault, including the Transverse Ranges, is still gliding northwestward with the Pacific Plate about 2 inches per year relative to the rest of the North American continent east of the San Andreas. The Transverse Ranges continue to rotate clockwise.

Earthquakes and Recent Deformation

Since the big Miocene blast, southern California has been one of the most tectonically active places in the world. Volcanic eruptions have punctuated the Mojave Desert in the last 25,000 years (Vignette 14), and earthquakes never cease. If we were feeling poetic, we might even describe southern California's geology as flamboyant.

Three major earthquakes in thirty years in the Mojave Desert portion of the Eastern California Shear Zone (an area of strike-slip deformation east of but parallel to the San Andreas fault) suggest that the Pacific Plate is preparing to take another bite out of the North American Plate. The 1992 M7.2 Landers earthquake, the 1999 M7.1 Hector Mine earthquake, and the 2019 M7.1 and M6.4 Ridgecrest earthquakes produced right-lateral strike-slip displacements of as much as 22 feet. The trend of their ground ruptures point up the east side of the Sierra Nevada, suggesting that the North America–Pacific Plate boundary will jump one day from the San Andreas fault (Vignette

19) into the Eastern California Shear Zone and Walker Lane, another strike-slip zone in western Nevada where already 25 percent of the shear between the Pacific and North American Plates is occurring.

Active and inactive faults underlie the fabric of southern California, but nuclear power plants have been built near them (Vignette 1), dams across them (Vignette 21), and cities and roads upon them (Vignette 19). We just hope that renewed slip doesn't occur during the lifetimes of these structures.

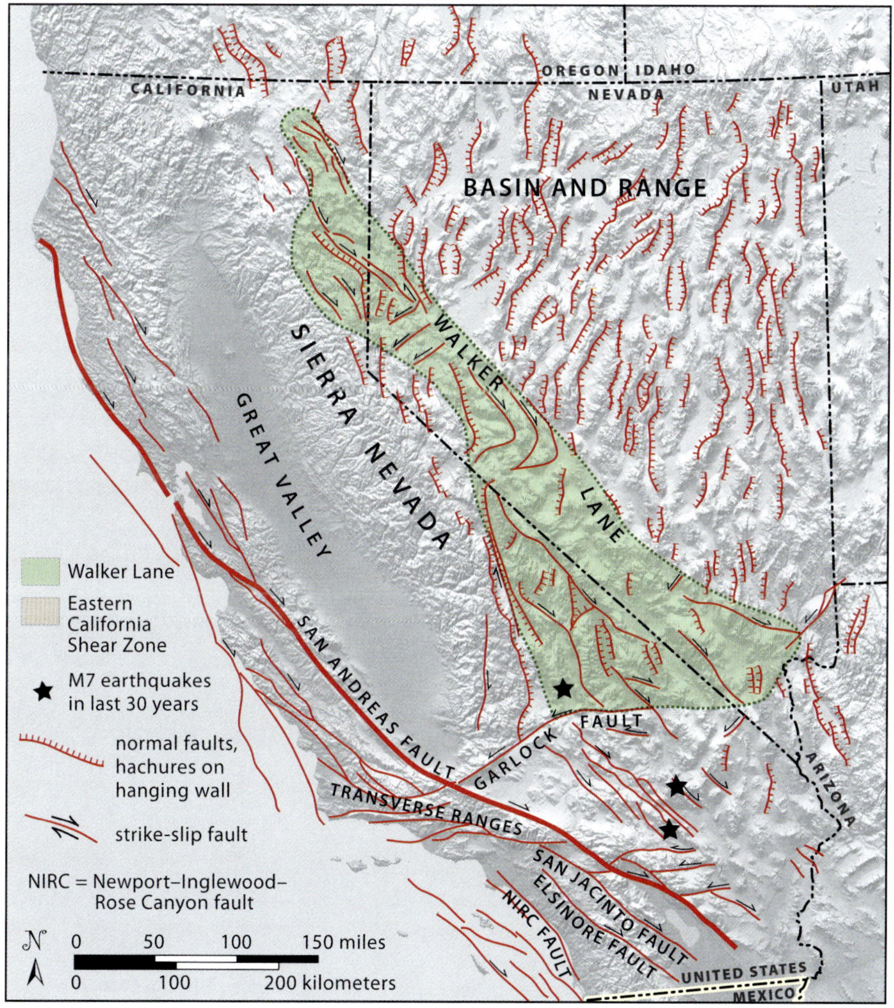

Recent major earthquakes in the Eastern California Shear Zone suggest that the boundary between the North American Plate and Pacific Plate may one day shift here from the San Andreas fault. —Modified from Henry, Faulds, and DePolo, 2007

Whenever an earthquake strikes California, or anywhere in the world, we soon learn what its magnitude was. But what exactly does it mean to be M4.0 or M7.5? Prior to 1932 when the first magnitude scale was developed, the size of an earthquake could only be reported subjectively by how it was perceived by humans and by the kind and amount of damage it caused. Eventually the Modified Mercalli Scale was devised. It rated earthquakes with Roman numerals on a scale from I to XII, where I is not felt by anyone and XII is total destruction. But this scheme was limited in being able to compare one earthquake with another. For example, a powerful earthquake could occur in an uninhabited area where no one was around to feel it, would cause no damage, and would therefore be assigned a Modified Mercalli Intensity of I. In contrast, a shallow, moderate earthquake beneath a heavily populated area with shabby construction would be felt by everyone, cause immense damage, and rate an intensity of XI or XII. Moreover, intensity depends upon location. At or near the epicenter (the point on the Earth's surface directly above the earthquake's origin), the damage could be intensity VIII or higher, whereas it decreases to zero with distance from the epicenter. Also, it took much effort and time to amass and compile intensity data. Nowadays, however, widespread use of the Internet allows intensity determinations to be automatically and rapidly managed through a US Geological Survey project aptly known as "Did you feel it?" (htts://earthquake.usgs.gov/data/dyfi/).

In 1932, Charles Richter and Beno Gutenberg of Caltech devised a numeric scale to compare the sizes of local earthquakes based on measurements taken from seismograph recordings. The Richter magnitude scale (M_L) became popular and widely used for many years, but it, too, had its limitations. It didn't work well for earthquakes greater than magnitude 5, nor for comparing sizes of distant earthquakes. It was based exclusively on local earthquake recordings made in California. Consequently, another Caltech seismologist, Hiroo Kanamori, developed the moment magnitude scale, M_w, which is a measure of the seismic energy radiated by elastic waves at the source of an earthquake. As such, it is a much better representation of the actual earthquake size. The numeric values of earthquakes in this book are moment magnitudes but are referred to simply as magnitude (M), thus: 1906 M7.9 San Francisco.

The moment magnitude scale is logarithmic; one integer step (e.g., from 2 to 3 or 6.5 to 7.5) corresponds to a tenfold increase in the amplitude of ground shaking and a 32-fold increase in the energy released. Two steps mean 10 x 10 = 100 times the ground shaking amplitude and 32 x 32 = 1024 times more energy release. Consider this: the 1964 M6.7 Northridge earthquake caused 57 deaths and damage estimated at roughly $50 billion in 2020 dollars. A magnitude 7.8 earthquake on the southern San Andreas fault, a scenario used by federal, state, and local officials for disaster planning (USGS "Shakeout2"), would shake the ground with over 10 times the amplitude and release nearly 40 times the energy.

All this tectonic activity makes southern California an exciting place to live. Deformation is so young and active that almost as fast as tectonic forces lift up areas, rivers, streams, and creeks do their best to erode them away. The erosion produces some of southern California's most visited landforms, including the gaps through mountains used by major highways.

The forces that cause earthquakes also fold rock and uplift land. If a tectonically produced fold forms a hill while a preexisting stream is flowing over it, and if the stream has sufficient abrasive power to cut down faster than the hill rises, then the stream may erode a steep-walled gorge across the rising fold. The stream will maintain its course across the fold as long as its rate of erosion is equal to or exceeds the rate of uplift. Such a stream is called an antecedent stream because it existed before the present topography was created. The erosional gap in the fold is called a water gap.

If, however, stream erosion is unable to keep up with uplift, then the stream will be deflected to a low area on the growing structure where it can cut a new water gap. The abandoned water gap is then called a wind gap. A long, laterally propagating fold may develop several wind gaps over time. One of the best examples of such a history is the eastward propagating Santa Monica Mountains. During their eastward growth over the last 4 million years, the mountains have progressively deflected the Los Angeles River around their east end, leaving several wind gaps behind from Sepulveda Pass eastward to the river's present course in the Elysian Park area. Other southern California examples of structures creased by a water gap include the Puente Hills (Vignette 9), cut by the San Gabriel River; Wheeler Ridge (Vignette 10), cut by Salt Creek; and the Ventura Avenue anticline (Vignette 4), cut by the Ventura River.

In addition to erosion by streams, other processes are also changing the landscape before our eyes. Sea cliffs are eroding at a rate of more than 1 foot per year (Vignette 2), and southern California's beaches are receding faster than natural replacement occurs (Vignette 5).

Wildfires are denuding forests and grasslands, making them more susceptible to landslides (Vignette 16), debris flows (Vignette 19), and floods. Stay tuned for the next geologic event to shape the future of southern California.

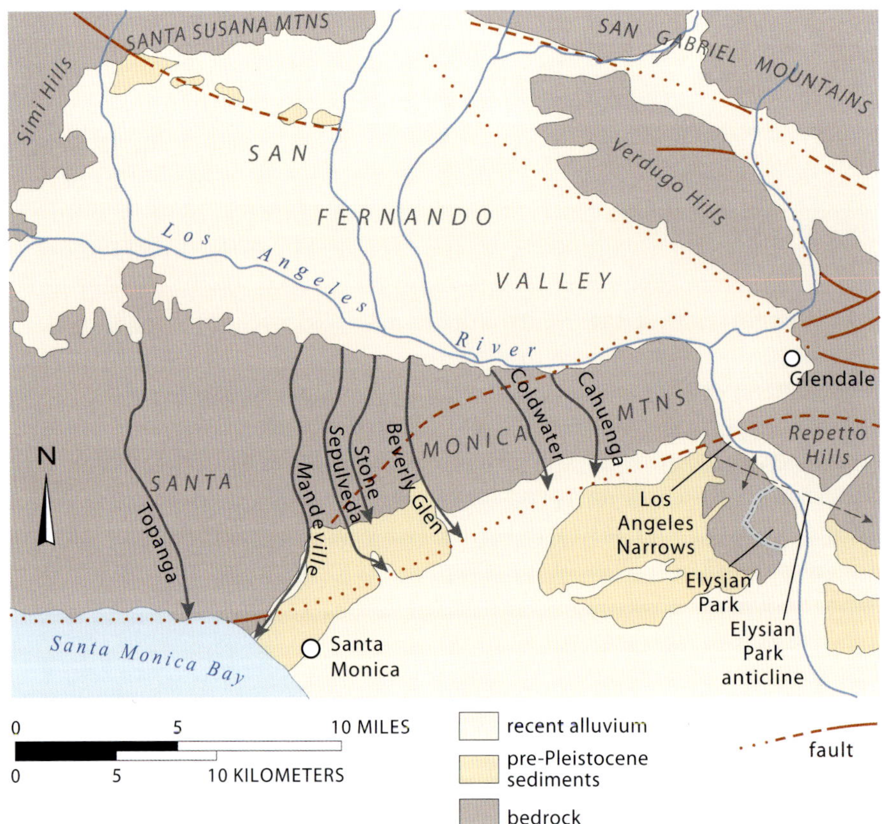

The Los Angeles River was progressively deflected eastward across and around the eastern Santa Monica Mountains during Pleistocene time, leaving deep wind gaps behind. I-405 follows Sepulveda Canyon, US 101 follows Cahuenga Pass, and I-5 follows the present course of the river at the east end of the range. —From Sylvester and Gans, 2016

VIGNETTE 1

SAN ONOFRE NUCLEAR GENERATING STATION AND THE CRISTIANITOS FAULT

Decommissioned and Inactive

SAN DIEGO COUNTY

People worry about nuclear power generating plants, particularly where geologic and seismic conditions at and near them are critical to plant safety. The San Onofre Nuclear Generating Station (SONGS) facility overlooks the Pacific Ocean from atop the San Onofre sea bluff, about 2.75 miles southeast of San Clemente and about 15 miles northwest of Oceanside. SONGS's location just 0.6 mile northwest of the Cristianitos fault raised serious questions about the plant's structural integrity before it was decommissioned in 2013. If slip were to occur on that fault, would it cause damage that would endanger the power plant?

The threat of seismic shaking to the structural integrity of nuclear power plants is very low because their structure and components are designed by earthquake engineers to withstand strong ground motion and remain functional during a maximum credible earthquake. In fact, power plant seismometers are designed to immediately and automatically shut down the power plant whenever strong ground motion is detected. Moreover, Dr. Larry Foulke, a distinguished earthquake engineer now retired from the Nuclear Regulatory Commission, Westinghouse Energy Systems, and the University of Pittsburgh, considers SONGS one of the most earthquake-resistant structures in California and says it is the safest place anyone could be during a major earthquake.

Nevertheless, people do worry that the Cristianitos fault might rupture someday and produce significant earthquake ground shaking. They fear that the nuclear power station might somehow be unintentionally compromised and release radioactive debris into the atmosphere. These concerns are understandable if the fault is still active. In general, faults with a history of fairly recent displacement are considered active; those that have not slipped for tens or hundreds of thousands of years are considered inactive. Is the Cristianitos fault active?

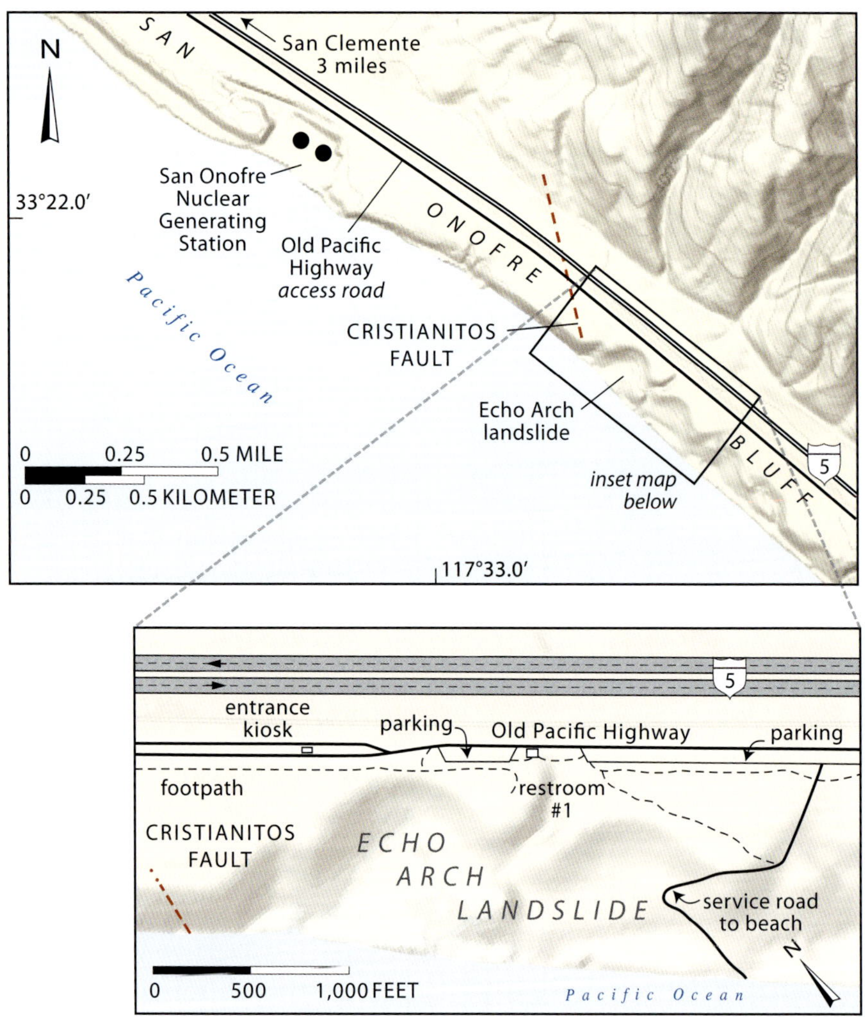

GETTING THERE

The Cristianitos fault exposure is located on the coastline between San Clemente and San Diego. To get there, exit the San Diego Freeway (Interstate 5) at Basilone Road and San Onofre (exit 71) and follow signs along San Onofre Road, first west and then south, to the San Onofre State Beach. This route follows the Old Pacific Highway past the nuclear plant. A state park entrance kiosk lies 0.8 mile south of the plant and 3 miles from Basilone exit. Drive 0.3 mile past the entrance kiosk through a fee parking area to a gated service road. Walk down this road all the way to the beach, and then walk north about one-third mile along the beach to the outcrop of San Mateo Sandstone. The total distance from where you parked your car is about a half mile. The climb out ascends 140 feet and may take 20 minutes.

The San Onofre Nuclear Generating Station, viewed from the southeast. Its foundation was excavated 20 feet down into the massive, white San Mateo Sandstone. A ribbed seawall protects the plant from wave erosion. Shingle beach gravel consists of San Onofre Breccia pebbles and sand, eroded from Dana Point 12 miles to the north. (33°22.13N, 117°33.28W)

Let us examine the geologic relationships in the field to evaluate that possibility.

The best way to view the Cristianitos fault and evidence for its activity is to walk down to the beach at the foot of the bluff at San Onofre State Beach via the gated service road to Trail No. 1. The service road goes across a curious hummocky bench surrounded by steep, 30- to 40-foot-high scarps of the Echo Arch landslide that indents the otherwise nearly vertical San Onofre Bluff. Echo Arch is the breakaway scarp, the exposed part of the slip surface on which the landslide moved. Be grateful for the slide for without it, we would have to rope our way down a 140-foot cliff to reach the beach.

Once on the beach, walk northwest approximately one-third mile to the north side of the Echo Arch landslide. A close look at the sedimentary strata exposed in the sea cliff and the adjacent landscape scarp reveals the details and geologic history of the Cristianitos fault, a major geologic structure. The fault extends inland a projected distance of about 26 miles on a bearing slightly west of north, crossing Ortega Highway (CA 74) and heading up Cañada Chiquita in Rancho Mission Viejo.

Aerial view of the north half of the vegetation-covered Echo Arch landslide. Reddish-brown Quaternary alluvial deposits exposed in the landslide scarp overlie both the white San Mateo Sandstone and the Cristianitos fault in the lower left corner of the image. —Photo © 2004 Kenneth and Gabrielle Adelman, California Coastal Records Project, www.Californiacoastline.org

Following the 1971 M6.7 San Fernando (Sylmar) earthquake, the California legislature passed the Alquist-Priolo Earthquake Fault Zoning Act of December 22, 1972. It set guidelines for construction of buildings for human habitation at or near earthquake-prone faults—such as the San Andreas, San Jacinto, Elsinore, Hayward, and other well-known faults in California. The bottom line: It is illegal to build a structure for human habitation across the trace of a known active fault. The legislation defined three categories of faults, as outlined in the table.

EARTHQUAKE FAULT NOMENCLATURE		
DESIGNATION	DESCRIPTION	CONSEQUENCE TO CONSTUCTION
ACTIVE* (Holocene)	Fault has ruptured the ground surface in the past 11,000 years.	No structure for human habitation (where people live) may be constructed across the trace of an active fault in the State of California. The generally accepted setback is 50 feet.
CAPABLE (for nuclear power siting purposes only)	Any nearby fault that has slipped within the past 35,000 years is considered capable of generating earthquake ground shaking that could damage a proposed nuclear power plant.	
POTENTIALLY ACTIVE (Quaternary)	Fault has ruptured the ground surface in the past 1.6 million years but not in the past 11,000 years.	In general, it is conventional wisdom to keep structures for human habitation at least 50 feet away from these faults.
INACTIVE	Fault has not ruptured the ground surface in the past 1.6 million years.	There are no generally accepted guidelines for inactive faults, beyond the generally accepted wisdom that such faults are not hazardous.

*In recent years there has been a move to replace the word *active* with the word *hazardous*. Either way, the same definitions apply. —Table excerpted from Hart and Bryant, 1997

Southern California has many faults—of the geologic variety. Some are active, but more are dead. Is Cristianitos fault active or dead? Take a close look at the fault. You will seldom have a better opportunity to put your finger right on one. Fault displacement produces earthquakes, and here is a beautiful fault, just 0.6 mile from a nuclear power plant. Should we have worried when the power plant was still active?

On the east side of the Cristianitos fault, gray, slide-prone Capistrano Formation underlies red-brown Quaternary alluvium. High tides and storm surf continually eat away at the toe of this sea cliff, creating never-ending rounds of erosion, slope instability, and landslides. The Echo Arch landslide occurred in the Capistrano Formation.

North of Echo Arch and west of the Cristianitos fault, durable, erosion-resistant San Mateo Sandstone forms an erosion-rilled, landslide-free, curvilinear coastline to, beneath, and north of the decommissioned nuclear power plant. The sandstone was deposited during Pliocene time, between 5 and 3 million years ago, whereas the Capistrano Formation is approximately 20 to 5 million years old. The difference in the ages of the two formations juxtaposed by the fault indicates that the younger San Mateo Sandstone has been displaced downward against the older Capistrano Formation. The slightly curved fault surface dips about 60 degrees west. Exposure to the ravages of weather and salt spray has obliterated any chance of seeing direct evidence of fault displacement, such as scratches and grooves or polishing of the fault surface, but if you could dig far enough into the cliff along the fault surface, there is little doubt you would expose such evidence.

The old wave-cut platform surface has absolutely no evidence of disruption or displacement by the Cristianitos fault. The marine boulder layer that lies directly upon the platform, as well as the 60 to 90 feet of overlying, dark red-brown alluvial deposits, are likewise unbroken. Isotopic ages of marine corals and shells in those unbroken overlying sediments are about 120,000 to 125,000 years old. For nuclear power plant siting, the Nuclear Regulatory Commission requires that the last movement on a nearby fault must not have occurred in the last 35,000 years. The Cristianitos fault is clearly older than that and so is not considered hazardous for siting a nuclear power plant.

How reliable is the 125,000- to 120,000-year age? As the shoreline receded westward across the now-emerged platform, it left behind a layer of nearshore deposits, including fragments of fossil seashells. Analysis of chemical changes among amino acids within the shells provides one way to approach the age problem, but a better age for these shells can be determined by analysis of the daughter products

The Cristianitos fault surface is visible behind the person in this sea cliff, 0.6 mile southeast of the nuclear power plant. The San Mateo Sandstone lies above and left of the inclined fault surface (dashed line); gray Capistrano Formation shale lies below and to the right. The fault does not break the layer of marine boulders and alluvial deposits that overlies the wave-cut platform (dotted line), indicating that displacement on the Cristianitos fault has not occurred since deposition of the alluvium. (33°21.782N, 117°32.717W)

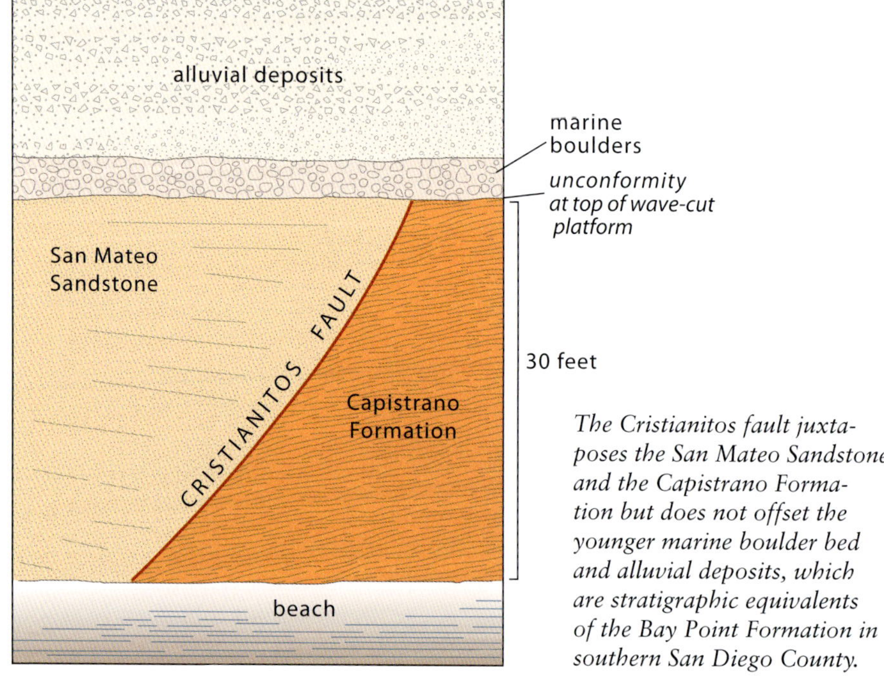

The Cristianitos fault juxtaposes the San Mateo Sandstone and the Capistrano Formation but does not offset the younger marine boulder bed and alluvial deposits, which are stratigraphic equivalents of the Bay Point Formation in southern San Diego County.

resulting from radioactive decay of the tiny amount of uranium within them. Those analyses yield an age of about 125,000 years. This procedure works especially well with fossil corals. Although rare on most southern California terraces, fossil coral fragments in the Nestor Terrace at San Diego yield consistent ages around 125,000 years. Is the wave-cut platform at San Onofre Bluff the same at the Nestor Terrace at San Diego? Yes: Careful geologic work confirms the correlation.

Minor faults of similar orientation cut the massive white San Mateo Sandstone on the west side of the main fault. This fractured zone is about 120 feet wide and is best seen from the beach around the corner from the north end of the Echo Arch landside. Many of these minor secondary faults have offsets of about 3 to 6 inches and dip about 15 degrees south, toward the main Cristianitos fault. Dark- to medium-gray Capistrano Formation layers, which weather grayish with some white layers of fine-grained sandstone, are severely buckled,

The 120-foot-high sea cliff 0.6 mile southeast of the nuclear plant displays, from bottom up, a rilled cliff of white San Mateo Sandstone that is smoothly truncated at its top by a horizontal wave-cut platform, overlain by a thin marine boulder layer, and topped by thick, reddish-brown nonmarine deposits. Several minor faults break the thin greenish-brown bed in the buff sandstone. The contact between the San Mateo Sandstone and the overlying alluvium is a major unconformity that represents a time when the surface of the sandstone was eroded by waves. The person stands at the base of the sea cliff for scale. (33°21.78N, 117°32.72W)

broken, and crumpled on the east side of the Cristianitos fault. What looks like bedding in these rocks is shear-banding parallel to the fault.

The smooth, horizontal upper surface of the white San Mateo Sandstone, the 120,000-year-old wave-cut platform, was eroded while it was below the high tide line. Sea waves are now cutting a similar platform as they eat away at the present sea cliff. Currents, swells, and breakers attack and erode the shoreline, undercutting the cliff until it caves in. The sea then clears the rubble away and again undercuts the cliff until it again caves. This continuing cycle creates a low, gently seaward-sloping surface—the wave-cut platform—that grows wider in the wake of the retreating sea cliff.

Sea waves cut such platforms below the level of high tide, but the one at the top of the San Mateo Sandstone is now 40 feet above high tide. Either the land rose or sea level dropped, or both. Relationships elsewhere suggest that sea level was about 20 feet higher when the presently emergent platform was cut. Between here and Dana Point, 14 miles up the coast, however, this platform rises nearly 90 feet above sea level. That slope means the land has risen at different rates at the two locations. So, about 20 feet of the emergent height of the platform at San Onofre is probably due to a drop in sea level, the other 20 feet to rise of the land.

The wave-cut platform on the sandstone is now buried by 60 to 90 feet of thick, dark-brown sedimentary deposits. Where did they come from? Imagine what would happen here tomorrow if sea level were to start falling, or if the land were rising. In either case, the shoreline would move seaward, exposing the modern wave-cut platform. The beach would move progressively seaward across it, leaving a thin layer of sand, beach stones, seashells, cans, bottles, sandals, and whatnot. Streams, slides, and mudflows would carry rocky, dirty, weathered debris from the inland hills behind the abandoned sea cliff onto the emerging platform. This debris would accumulate initially on the platform's landward edge, building an apron sloping seaward. As this apron thickened and steepened, it would also grow seaward because streams could carry debris across its sloping surface to the outer edge. All of that debris would be the equivalent of the boulders and brown gravel that now cover the old wave-cut platform.

It might take the better part of 100,000 years to accumulate the 60 to 90 feet of debris exposed in the present bluff. Once sea level stabilized, waves would again erode the land, carving a new wave-cut platform and creating a new sea cliff, revealing the newly deposited debris we see exposed in it. This sequence of events has happened repeatedly along the southern California coast, creating a succession of emergent wave-cut platforms and abandoned sea cliffs, which, together with the nearly flat expanses of old seafloor beyond them, are called marine terraces. Four easily recognized marine terraces border the coast between San Onofre Bluff and Oceanside like a flight of giant steps rising from the beach inland into the hills. As many as thirteen terraces notch the Palos Verdes Hills, 65 miles to the north.

Although displacement hasn't occurred on the Cristianitos fault since 60 to 90 feet of alluvium were deposited upon the wave-cut platform, faults offshore may be more of a threat to the San Onofre Nuclear Generating Station (SONGS) and other coastal structures. Even though no one has actually seen these subsea faults, buried as they are under mud on the ocean floor, geophysical evidence indicates that a southern extension of the Newport-Inglewood fault, which caused the 1933 M6.4 Long Beach earthquake, connects with the Rose Canyon fault, an active fault in the San Diego area, by an offshore fault only a few miles from the nuclear generating station.

What are the lessons of San Onofre Bluff? Good exposures of simple geologic relationships can provide information of real value to society. Fears, real and imagined, about the hazard to SONGS posed by an earthquake on the Cristianitos fault can be assuaged when we are able to see and read the geologic history as recorded in the sea cliffs.

While this book was in press, the California Coastal Commission gave Southern California Edison Company permission to dismantle the decommissioned San Onofre Nuclear Generating Station. Work was anticipated to commence in 2020 and be completed in 2028. The distinctive twin domes, rising 200 feet between I-5 and the Pacific Ocean, are expected to come down in mid- to late 2025. The substation without transformers, a security building, the seawall, a walkway connecting two beaches north and south of the plant, and a switchyard with power lines will stay put.

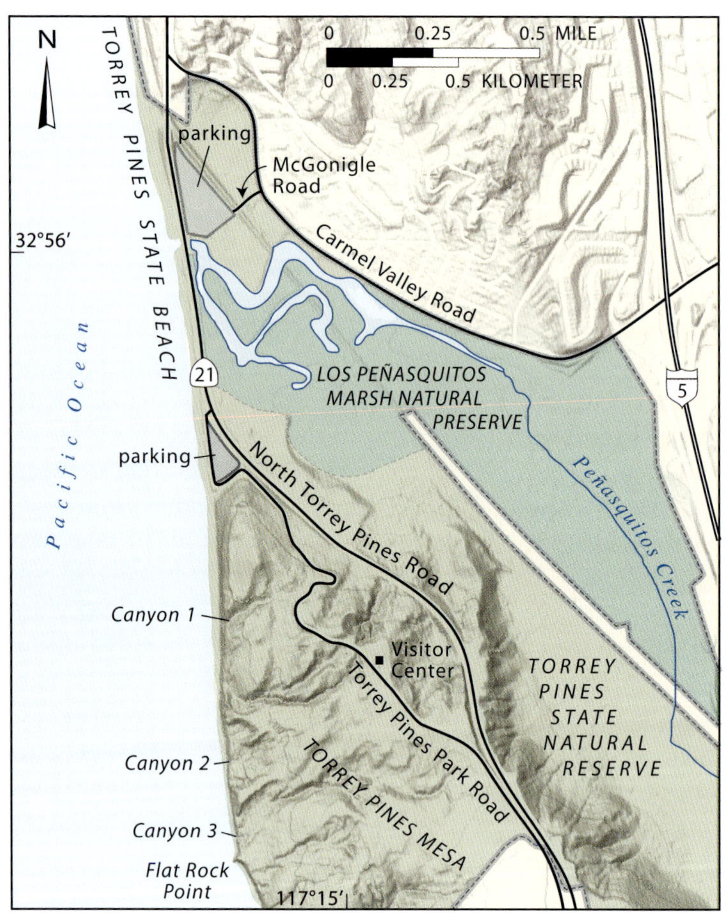

GETTING THERE

Take Carmel Valley Road (exit 33) from Interstate 5. Drive west 1.5 miles to County Road S-21 (Old US 101). Turn south and drive 0.8 mile across the bridge over the mouth of Peñasquitos Creek to the Torrey Pines State Natural Reserve fee parking lot, public restrooms, and beach access.

VIGNETTE 2

TORREY PINES STATE NATURAL RESERVE

A Migrating Shoreline

SAN DIEGO COUNTY

Torrey Pines State Natural Reserve occupies an attractive section of the San Diego coastline and a unique botanical niche in California. Botanists consider its pine trees, *Pinus torreyana*, ecological relics of cooler and wetter conditions of the last ice age, between 75,000 and 11,000 years ago, when they were widely distributed along the coast. Under today's warmer and drier conditions, only small clumps of Torrey pines survive in favored spots, such as on Torrey Pines Mesa and a few orphans on Santa Rosa Island in Channel Islands National Park. But we leave these topics to the botanists and ecologists. Our concern is more with the eons of geologic history revealed in the layers of sedimentary rocks exposed in the spectacular sea cliffs.

The bluffs at Torrey Pines rise abruptly 100 to 200 feet above the sandy beach. These impressive cliffs, as well as Torrey Pines Mesa above and to the east, are bounded on the northeast by the wide, flat, swampy bed of Peñasquitos Creek in Soledad Valley. The valley is an artifact of the last glacial maximum 20,000 years ago, when enormous glaciers in northern latitudes withdrew enough water from the ocean to reduce sea level about 300 feet. During that time, a stream eroded the floor of Soledad Valley down to meet the lower sea level. As the melting glaciers returned their water to the oceans, sea level rose and flooded the lower part of the valley, creating an estuary like those farther north toward Oceanside. Streams flowing into the Soledad estuary filled it with sediment, transforming it into a lagoon and tidal wetlands, its floor hardly above sea level. Where North Torrey Pines Road (Old US 101) crosses the mouth of Soledad Valley, one would have to drill through about 100 feet of sediment to reach the old valley bottom.

After you find a place to park, let's consider the beach as you walk to the sea cliffs. Most southern California beaches change form dramatically with tides, seasons, and storms, and Torrey Pines beach is no exception. San Diego's beaches are usually wide and full of sand and sunbathers during summer months—a boon to businesses, tourists, and locals alike to be sure, but surfers must wait until winter for better surfing waves to arrive from storms originating in the Gulf of Alaska that send short-period swells south. Waves from these swells break closer to the beach, allowing higher energy backwash to carry sand away from the shore. Long-period summer swells arrive at this coast from distant storms generated in the southwest Pacific Ocean. They generate waves that "feel" the bottom in deeper water more than do short-period waves, so they break farther offshore and generate a low-energy forward swash that carries sand onto the beach. Thus, sandy beaches usually erode in winter, exposing well-rounded, gray and varicolored pebbles and rocks, especially at low tide, that you rarely see in summer when waves return sand to the beach. We'll return to these pebbles and rocks on our return walk to the parking area.

Stand near the north end of the nearly vertical bluffs (32°55.53N, 117°15.57W), preferably at low tide, but near the water's edge, so you can see their full height and majesty and how they expose nearly horizontal layers of sedimentary strata. Rock layers in the lower part of the cliff are 49 to 47 million years old, from middle Eocene time. In

View south of the thin-bedded Delmar Formation (1), which underlies massive tan and reddish-brown Torrey Sandstone (2). The uppermost dark, reddish-brown cliffs are the Late Pleistocene Bay Point Formation (3). The three formations are separated from one another by unconformities (dashed lines). (32°55.273N, 117°15.540W)

California, rocks that old are commonly folded into inclined layers—anticlines and synclines—but these nearly horizontal layers testify to the tectonic stability of this small part of the Earth's crust—the San Diego block—in an otherwise dynamic and earthquake-prone California.

Notice that the shape of the sea cliff profile is nearly vertical from the base to within about 50 feet of the top. The Bay Point Formation, a relatively young, more easily erodible sequence of sedimentary

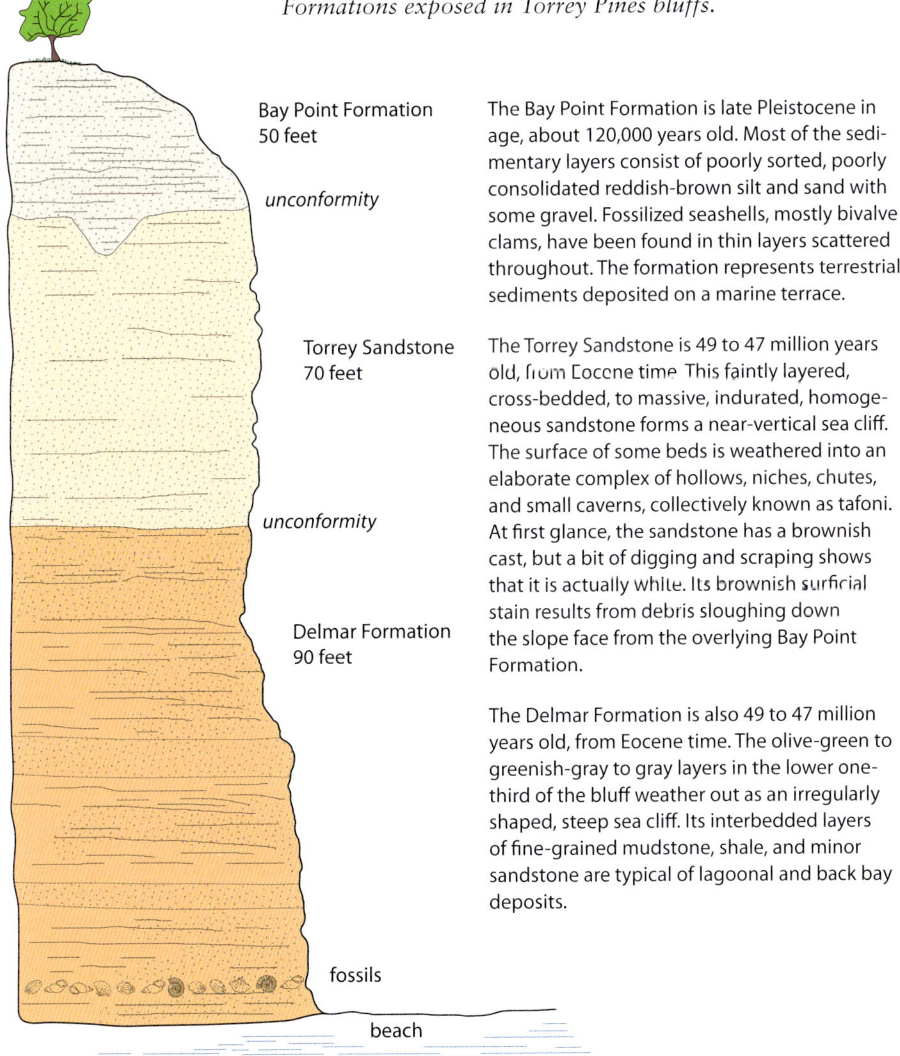

Formations exposed in Torrey Pines bluffs.

layers, forms the upper, somewhat rounded brush and grass covered slope. Nearly vertical rills and gullies give the formation's slope the appearance of miniature badlands.

Let's take a closer look at the sedimentary rocks that are oldest and closest to us, those at the base of the Torrey Pines sea cliff. These nearly horizontal, seaward edges of the Delmar Formation and Torrey Sandstone are widespread sedimentary deposits that extend several miles inland. Normally, sediments are laid down over broad areas in flat-lying, successive layers, one on top of the other—like stacks of pancakes. That makes the bottom layer the oldest, the top layer the youngest. But not quite so fast; let's take a closer look at the sequence of deposition for these two formations at Torrey Pines.

Although the Torrey Sandstone is correctly mapped above the Delmar Formation, the two formations were actually deposited at the same time and intertongue throughout. In any one outcrop, you may see a lens of Torrey Sandstone sticking into the Delmar Formation, or the reverse, a tongue of Delmar Formation sticking into the Torrey Sandstone. Without seeing the big picture, however, small outcrops can be confusing as to which formation is which. Here at Torrey Pines State Natural Reserve, Torrey Sandstone dominates the upper part of the middle Eocene section in the sea cliffs, and the Delmar Formation dominates the lower part.

About 50 million years ago, this coast was a low plain, and shallow water stretched a considerable distance offshore. Waves break where they get into shallow water, so the line of breakers was far offshore—it would have been a good surfing spot. The breakers scoured sand from the seafloor and piled it into a broad sandbar. The bar may have risen above sea level to become an offshore barrier island, enclosing a shallow lagoon between itself and the coast. Finer mud and silt accumulated in the calm lagoon. That ancient coast probably looked

EARLY HUMANS IN SOUTHERN CALIFORNIA

After about 11,000 years ago, aboriginal people occupied coastal California, where the weather was mild and seafood was abundant. Every generation or so, their camps had to be moved inland to keep ahead of rising seas. Divers have found discontinuous layers of submerged pottery, stone tools, metates, and grinding stones that parallel the shore, indicating times when sea level stabilized for a while. These relics of human occupation attest to lowered sea levels in the not-too-distant geologic past.

much like the modern Texas-Louisiana coast, with its barrier islands and lagoons. Inlets through the barrier island let water pass in and out of the lagoon as the tide rose and fell.

Occasional storm waves, especially at high tide, swept over the barrier island, spreading a layer of sand across the mud in the lagoon. Flooding streams from the mainland also carried sand into the lagoon. Plants and animals generally abound in lagoons, so the lagoonal sediments are richly fossiliferous.

The Torrey Sandstone looks like a barrier island or offshore sandbar deposit and the Delmar Formation like lagoonal muds. But how did the sand bank come to lie on top of the lagoonal mud? As long as sea level remains stable, a barrier island and its lagoon remain reasonably fixed in position, but if sea level rises, or the land sinks, the shoreline migrates inland.

As waves attack the seaward side of the barrier island, they wash sand into the lagoon, at least along its seaward edge, probably accounting for the massive sand beds in the upper part of the Delmar Formation. This migration of bar and lagoon is so slow that a considerable thickness of lagoonal deposits may accumulate before the bar sands arrive on the scene. In due time, the barrier island moves landward across the old lagoon, burying the lagoonal muds under sand.

Most of us have seen the large contraptions that spread concrete or asphalt for a modern highway. The paving machine lays down a ribbon of material of uniform thickness from one end of the project to the other. The farther part is older, the nearer part younger. Now imagine two paving machines many miles wide, one following the other, and each laying down a sheet of sediment across a smooth plain. That, basically, is what happened in the Torrey Pines area 49 to 47 million years ago.

First, either rising sea level or sinking land allowed the shoreline to move several miles inland, and the machine paving the transgressing barrier island laid a sheet of sand over the lagoonal mud. Then sea level dropped or perhaps the land rose. As the shoreline moved seaward, the machine paving the lagoon spread mud over the layer of sand the barrier had left as it moved seaward. Then the sand paving machine moved inland again. Few places in North America demonstrate the inland movement of this paving machine mechanism better than the cliffs at Torrey Pines.

The unconformable contact between the Torrey Sandstone and the overlying Bay Point Formation represents a major gap in the rock record, a long passage of time during which sedimentary layers either were not deposited or were deposited but subsequently eroded away.

In the San Diego region this erosion surface is known as the Nestor Terrace, which has been dated at 125,000 years old. Near the north end of the sea cliffs, the contact is relatively smooth and level; farther south, deep gullies eroded into and even through the Torrey Sandstone and are now filled with much younger Bay Point Formation rocks. Corals found near the base of the Bay Point Formation have been dated by uranium-series methods at approximately 120,000 years before present.

A Walk along the Base of the Bluffs to Flat Rock

CAUTION: A danger sign at the access stairs warns beachgoers to stay back from the cliffs, because they are marginally stable at best and may fail at any moment. Approach the east edge of the beach with abundant caution. This is a State Natural Reserve! Leave your rock picking tools in your car but bring your camera to record splendid memories of one of the finest exposures of Quaternary and Eocene sediments along the San Diego County coastline.

Keep track of the major canyon mouths as you stroll south along the beach from the beach access stairs opposite the parking-lot restrooms (32°55.60N, 117°15.56W). Canyon No. 3 (32°54.895N, 117°15.51W) is about 1 mile south of the beach access stairs and about 100 yards north of Flat Rock (32°54.83N, 117°15.53W).

Walk to the base of the bluff for a closer look at the lower part of the Delmar Formation and its blocks of rock that have fallen to the beach from layers high in the sequence. Look up to see the overlying, lighter-colored Torrey Sandstone and its cavernous weathering (tafoni).

As you walk along the cliff base a few hundred feet south of the parking area, massive beds of gray Delmar Formation rock form prominent ledges, and blocks fallen from them litter the cliff base. Look closely: Fragments of fossils are so abundant in some of the blocks that geologists call the rock "fossil hash." Many of the shell fragments are gray, thick, and finely laminated fragments of oysters. The thick shell structure of oysters allows them to live near the shore where water turbulence would destroy a clam with a thinner shell structure. Thinner, more fragile pieces of clamshells—like razor clams—are found in quiet water deposits. Layers of mudstone contain elegantly coiled, marine snail shells. The farther south you walk, the more fossils you see in the Delmar Formation and the better their preservation.

Walk south along the cliffs as far as Canyon No. 3 and Flat Rock, known to locals as Bathtub Rock. Its name derives from a hole, about

Broken oyster shells in a gray Delmar Formation sandstone bed. Pencil is 5.5 inches long. (32°54.85N, 117°15.50W)

4 feet by 7 feet across and about 2 to 3 feet deep, dug into the rock surface in the early twentieth century by a Welsh miner looking for coal. Waves splash water up onto the flat surface, and some of it remains in the "bathtub" for beachgoers to sit in and enjoy with libation at sunset. The rock consists of carbonate-cemented Delmar Formation. The cement is derived from the abundance of oyster shells that are clearly visible in this dark-gray, fine-grained rock. Boring clams (*Penitella penita)* made the little dimples and pits in the rock surface.

In the cliff adjacent to Flat Rock is a nicely exposed angular unconformity within the Delmar Formation. An angular unconformity is a surface upon which sedimentary layers are at an angle to each other instead of being parallel. The angular discordance here is about 10 degrees, between a horizontal massive sandstone bed and the tilted, thinner sandstone and shale layers beneath it. The sandstone and shale layers, one of which is full of broken oyster fossils, are inclined because they were deposited on the side of a shallow scour channel. Erosion beveled a flat surface across the tilted shale beds before the sandstone was laid down upon them.

On your way back to the parking lot, take an up-close and personal look at some magnificent cross beds in Torrey Sandstone in the base of the cliff near the mouth of Canyon 2. The cross beds are the numerous crisscrossing layers of sand in a pebbly, fine- to medium-grained sandstone. Individual cross beds dip on the order of

The dashed line marks the unconformable contact between the dark-gray, fossiliferous mudstone of the Delmar Formation and the overlying white and reddish-brown sandstone beds. The sandstone beds are 1.5 feet thick. (32°54.85N, 117°15.50W)

Sedimentary structures like these cross beds in sandstone typically form in offshore bars and reflect the variable nature of waves and currents in and near a surf zone. Staff is 4.5 feet long. (32°55.131N, 117°15.552W)

about 5 to 10 degrees below horizontal. The cross-bedded sandstone is overlain by a 1-foot-thick bed of black sand containing clots of brown sand. These clots are chunks of the older (stratigraphically lower) cross beds that were ripped up and transported by means of submarine slumping to this higher stratigraphic location in the Torrey Sandstone.

Beach Pebbles from Mexico

As you return to the parking area, notice well-rounded pebbles littering the beach, especially in winter. They are smooth, hard, mostly volcanic rocks that resist both the mechanical wear of the waves and the chemical corrosion of seawater. They originated from vast Eocene volcanic fields in the northwestern Mexican state of Sonora, where mechanical erosion broke the rocks into smaller and smaller pieces, which were then transported downslope to nearby rivers. These Eocene rivers flowed west and carried the volcanic clasts into the western San Diego area before the Peninsular Ranges existed. They were deposited along the coast in vast alluvial fans, the most prominent being named the Stadium Conglomerate, formerly known as the Poway Conglomerate, and extensively exposed where they line the cliffs in San Diego's Mission Valley. Over time these clasts were again eroded, some being deposited in the Lindavista Formation, which is exposed upon Torrey Pines Mesa but not in the sea cliffs.

Now fast forward to the recent geologic past—the last million or so years—and present. Erosion of the Stadium Conglomerate and the Lindavista Formation has supplied gravel to local rivers and streams, which was then carried to the present-day shoreline. Today, we can see these remnants of Mexican volcanoes in the back-beach after wintertime storms have washed summer sand out to sea. Clearly these pebbles have survived the abrasion mill several times. Many are little elongate disks, quite different from the more equidimensional spherical shape typical of stream-rounded pebbles. Ocean waves slide and roll them on the sand beach, wearing them to that common beach pebble shape.

With the passage of geologic time, these smoothed, rounded, and very durable volcanic beach pebbles may again become part of the geologic record. As changing sea levels alter this coastline, deposits preserved from this sandy beach and the distant estuaries may provide some far-in-the-future alien culture glimpses of the landscape we are experiencing today.

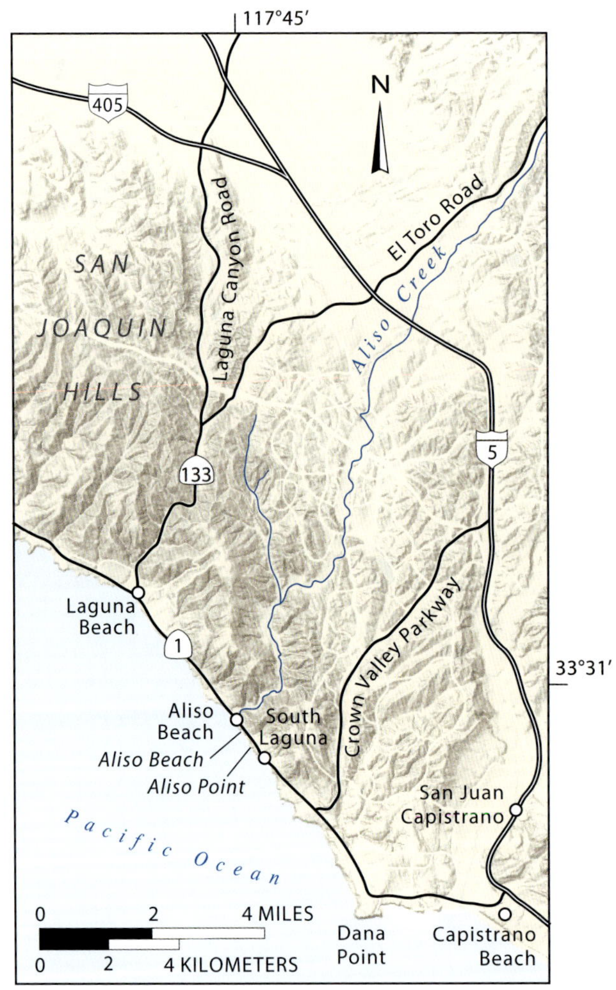

GETTING THERE

Aliso Beach is on CA 1, 9 miles south of Corona Del Mar, 3 miles south of Laguna Beach, or about 5 miles north of Capistrano Beach. The beach is accessible from a fee parking lot.

VIGNETTE 3

SAN ONOFRE BRECCIA AT ALISO BEACH COUNTY PARK

A Vanished Landmass

ORANGE COUNTY

Aliso Beach, one of the nicest of the many attractive cove beaches on the Laguna coast, has pale, coarse sand with a scattering of dark grains. Aliso Creek nourishes this beach with an ample supply of sand carried from the adjacent San Joaquin Hills and nearby Santa Ana Mountains. But its sand is not why we suggest you visit Aliso Beach Park in South Laguna. This county park has some of the best and most accessible exposures of the San Onofre Breccia in southern California.

The San Onofre Breccia (pronounced "bretch'-uh") is one of the most unusual sedimentary formations in southern California. The late A. O. Woodford, a Pomona College professor who lived to one hundred years and came to work regularly in a suit and tie well into his nineties, first described the formation in 1925 in a classic scientific paper. Like other breccias, the San Onofre Breccia is a mixture of sharply angular rock fragments embedded in a matrix of finer particles. The fragments range from fine gravel to blocks the size of a bus and consist of several rock types.

San Onofre Breccia is a remarkably unlikely collection of rock debris shed from an elevated landmass that stood somewhere nearby. Weathering and erosion, including rockslides, rockfalls, landslides, and debris flows, dumped these rock fragments into streams and rivers that carried them relatively short distances from their highland source and deposited them in the San Onofre Breccia basin. That happened between 20 and 15 million years ago (early to middle Miocene time), when the Laguna coast was quite different from what we see today. That highland is gone now, but its angular fragments are preserved in the breccia. Fossils of sea creatures within parts of the breccia indicate that some of the debris accumulated in the ocean. Most of it contains no such fossils, however, suggesting that the bulk of the breccia accumulated on land above the reach of ocean tides.

Let's inspect the breccia for ourselves to see what the rock fragments look like and what type of landmass they might have been eroded from. From the south end of the Aliso Beach parking area (33°30.50N, 117°45.07W), walk about 300 feet southeast down the beach to the first projecting cliff, which exposes sandstone within the breccia. Notice all the nearly vertical, irregular hollows in the higher part of the cliff. Such cavernous weathering is common in well-cemented, massive sandstone exposed in a sea cliff, but rarely will you see it in the same formation where exposed inland. The vertical hollows were formed by waves carrying beach sand up and down the face of the sea cliff, like sandpaper, during storm surges. Some parts of the sandstone weather and erode faster because they are less well-cemented than the more resistant parts. Stand back and notice that the gravelly layers underlying the sandstone dip gently southwest, toward the ocean.

Behind the cavernous weathering outcrop is a reentrant in the cliff face with a brown-stained, planar band cutting through the breccia, dipping to the south about 50 degrees. Beds on opposite sides of this band do not match, yet they are both part of the same geologic formation. In this case, different parts of the breccia have been brought

Bedding dips gently to the right in this sandy block of San Onofre Breccia that exhibits cavernous weathering. Two-foot-high traffic cone for scale. (33°30.47N, 117°45.06W)

An angled fault (between the arrows at left) dips about 50 degrees south and separates the coarse, cavernous-weathered brown breccia and sandstone (left) from the gray sandstone (right). A nearly vertical fault on the right side of the reentrant displaces the gray rock layers a couple of feet down to the west. Can you match the beds from one side of the vertical fault to the other? (33°30.47N, 117°45.06W)

together—one body slid past the other along a fault. This narrow reentrant in the cliff formed because two faults, on either side of the reentrant, are zones of weakness and are relatively more erodible than the rocks they cut.

Walk another 100 feet to the next projecting cliff where several large blocks of rock have fallen along its base. The various layers of breccia here contain fragments as small as pebbles and some as large as boulders. All are embedded in a sandy groundmass of finer rock particles.

Look carefully at individual rock fragments. Most are angular, with sharp edges, although some edges have been blunted by impact and abrasion during transport. A few of them are well rounded, suggesting that they traveled from a more distant source. Fragments vary in color, texture, and composition, which means that they came from an area made up of various rock types. The dominant rock type in the San Onofre Breccia is a dark gray schist full of spots that make little

This cliff of San Onofre Breccia exhibits coarse pebbly sandstone with interbeds of angular cobbles and boulders. Staff is 4.5 feet long. (33°30.46N, 117°45.05W)

bumps on the rock's surface. Schist is a metamorphic rock that has a distinctly laminated structure, as though it were a stack of thin sheets; this is foliation (from the Latin *folium*, meaning "leaf").

Fragments of greenish schist, a metamorphosed volcanic rock, are abundant. Chunks of white quartz are probably pieces broken from quartz veins. Tan to light-gray fragments of igneous rocks are only weakly foliated. The platy shape of most of the fragments in the breccia is best explained by their tendency to break parallel to their foliated texture.

Look here and farther along the cliffside for platy fragments of a fine-grained, shiny rock, that, in good light, has a distinctly lavender or bluish color. Known as glaucophane schist, this is easily the most distinctive rock among the San Onofre Breccia fragments. Glaucophane, an uncommon variant of the common mineral amphibole, forms at crustal depths of 20 to 45 miles, but most places in the Earth are too hot for this mineral to form. The only place cool enough at that depth is in a subduction zone, where cold seafloor rocks descend into the mantle. Thus, glaucophane, so easy to recognize in the field, is a sure sign that you're looking at rocks that went down a subduction zone, were metamorphosed there under high pressure and relatively low temperature, and then were brought back to the surface by tectonic forces and erosion. In southern California, glaucophane is

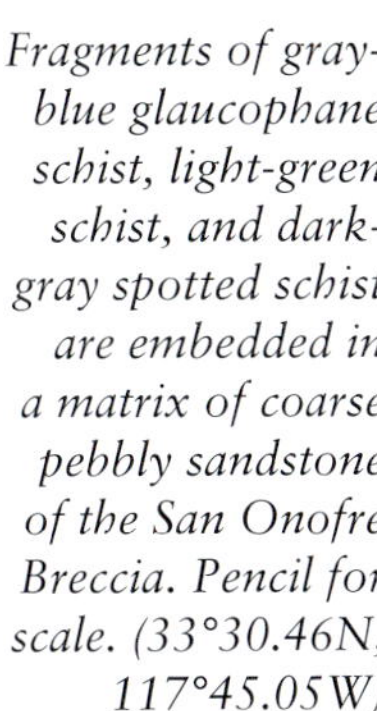

Fragments of gray-blue glaucophane schist, light-green schist, and dark-gray spotted schist are embedded in a matrix of coarse pebbly sandstone of the San Onofre Breccia. Pencil for scale. (33°30.46N, 117°45.05W)

found only in the Franciscan Formation, an assemblage of rocks that was drawn down into a subduction zone that existed until about 25 million years ago along what is now the southern California coast.

Continue your stroll about another 350 feet along the shore to Aliso Point, a ragged cliff with a prominent knob, or sea stack, standing about 50 feet shoreward from the sea cliff. The rocks in the knob and in outcrops in the adjacent sea cliff contain several types of veins.

One group of veins is white, homogeneous calcite—that is, calcium carbonate. White veins are common in rocks, and they are almost always either quartz or calcite. A quick field test to see whether it is a calcite or quartz vein is to scratch it with a knife or other piece of iron or steel. The tip of the corkscrew on a pocketknife makes a particularly fine rock scratcher; it will scratch calcite but not quartz.

Once the breccia deposits were firm enough to fracture, circulating groundwater deposited calcite in long, planar cracks. On the face of the first outcrop inland from the sea stack, the whole side of one vein has been laid bare, so it looks as if the exposure were coated with calcite.

White calcite veins cut this greenschist boulder in the San Onofre Breccia at Aliso Point. Sunglasses for scale. (33°30.42N, 117°45.04W)

This long, thin sandstone dike (between arrows) cuts the San Onofre Breccia at Aliso Point. Staff is 4.5 feet long. (33°30.42N, 117°45.04W)

A 30-foot-high sea cliff about 500 feet southeast of Aliso Point exposes extremely coarse San Onofre Breccia. The largest boulders are about 3 feet across. (33°30.38N, 117°44.95W)

The other veins are tan or gray, and generally thicker and more irregular than the calcite veinlets. They consist of sandy material like the matrix of the breccia, and some are thick enough to include small rock fragments. One vein on the south face of the Aliso Point sea stack looks like concrete grout injected into the breccia. These *clastic dikes* formed by injection of sand into fractures.

A dike is a tabular rock body that has forcibly intruded, or passively fills, an open fracture in a rock. *Clastic* simply means it consists of broken rock or mineral fragments. These clastic dikes formed when soft, deeply buried, fine-grained sediment liquefied and was injected under high pressure into overlying fractured rock, perhaps during strong ground shaking caused by an earthquake.

Venture around Aliso Point and about 500 feet into the next cove beach to see huge boulders, some as large as a washing machine, in the San Onofre Breccia exposed in the sea cliff. Here, the gravelly breccia dips about 17 degrees southward.

Search for the Source of the San Onofre Breccia

By now, you are probably coming to the realization that the San Onofre Breccia is an exceptional and unusual deposit. Its rocks are the age and kind that originated as seafloor sediments and lava flows subducted beneath a continent, metamorphosed, and finally uplifted to the surface. But where did that happen, and where are those parent rocks today? No parent rock that could have been its source has been found to the east or anywhere nearby. The right kind of parent rock—the Franciscan Formation—does exist in a few places west and north of Aliso Beach, such as on Catalina Island and in a small exposure on the Palos Verdes Peninsula, 40 miles to the north.

In 1925, Professor Woodford postulated that perhaps a high, rugged landmass stood just 1 to 2 miles off the present shore during late Oligocene to Miocene time, some 20 to 15 million years ago. Geophysical evidence indicates that major faults, along which a landmass may have risen, lie offshore, and thus, any east-facing fault scarps (cliffs) along these faults could have shed the sort of angular rock fragments that became the San Onofre Breccia.

So where is that landmass today? More fundamentally, how did it come to exist in the first place? The short answer is that the San Onofre Breccia represents the fragmented upper part of seafloor rocks previously subducted beneath the western edge of North America and now exposed in the region offshore of southern California. Initially an elongate slab of the North American Plate, comprising southern California and Baja California, was kidnapped by the Pacific Plate about 20 million years ago. In the process, the slab delaminated from the subducted rocks underneath and rotated in the way that a rug might be rotated on a floor. Seafloor rocks previously covered by the rotated piece were thus exposed, uplifted, and eroded, yielding rocks and sediments that became the San Onofre Breccia.

Numerous faults cut southern California, many with horizontal displacements measured in tens to hundreds of miles. Some of these faults also have a component of vertical displacement that could have uplifted the San Onofre Breccia source rocks. Most of the major faults of southern California have right-lateral displacement—if you look across the fault, the opposite side slips to your right. If an offshore right-lateral strike-slip fault carried away the source rocks for the San Onofre Breccia, then they must now lie somewhere to the northwest, up the coast—and they do.

One major right-slip fault, only 1 mile or so offshore, is the active Newport-Inglewood fault. It stretches southeastward from West Los Angeles near UCLA, through Long Beach, thence offshore, where geophysical and seismic investigations show that it continues past Aliso Beach and on to La Jolla, where it comes back onshore as the Rose Canyon fault and continues south through the San Diego area. The same investigations found rocks on the southwest side of the fault, both in offshore drilling and in deep onshore wells in the Long Beach area, that are the right type to be a source for the San Onofre Breccia.

And how about the Palos Verdes Hills? They have major faults along their inland side, and they contain a small exposure of metamorphic rocks of the right kind to have been the source of San Onofre Breccia at Aliso Beach. Much more of that rock may exist there but is now buried under younger sedimentary deposits. This required distance of displacement, roughly 40 miles, is sobering, but other southern California right-slip faults have had even greater displacements within the last 20 million years.

Aliso Beach is not the only place where San Onofre Breccia is exposed, but it is one of the best. The formation crops out in sea cliffs and hillsides all along the Laguna coast, especially at Dana Point, as well as in the hills inland from San Onofre State Beach, south of San Clemente. In Oceanside, the southernmost exposures of San Onofre Breccia can be seen in the walls of the San Luis Rey River channel and in roadcuts on the north side of Oceanside Boulevard, west of Interstate 5. Breccia beds extend inland a few miles, becoming finer grained with increasing distance from the coast, which is consistent with their postulated offshore source.

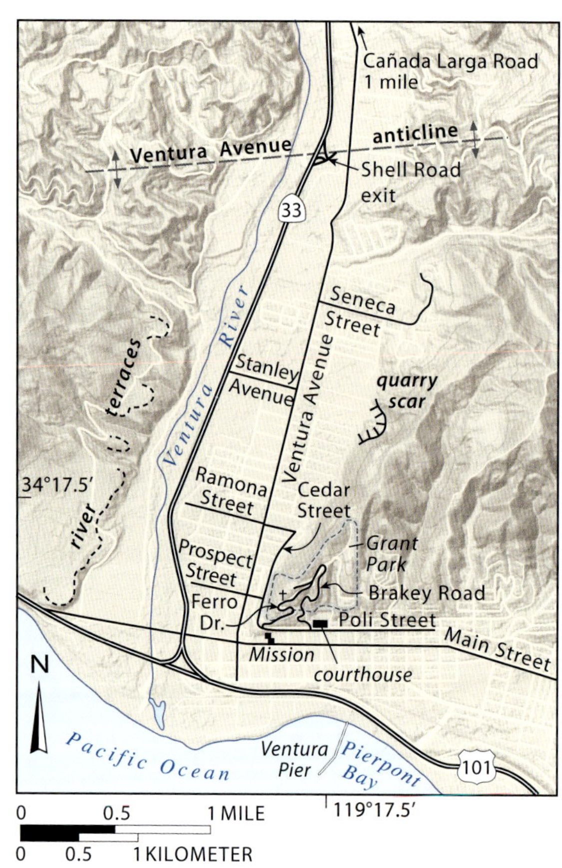

GETTING THERE

To see Ventura Avenue anticline, we'll first head north on CA 33 by taking exit 70B off US 101 in Ventura. Then, take Cañada Larga Road exit to head back south on Ventura Avenue. We'll end at the Grant Park regional vista overlooking Ventura.

VIGNETTE 4

VENTURA AVENUE ANTICLINE

An Oily, Growing Fold

VENTURA COUNTY

The kaleidoscopic landscapes of southern California are still forming and changing as the Earth's crust moves. These crustal movements—associated with our unwelcome earthquakes—are largely responsible for the varied landscapes. Most southern Californians know that a fault is a fracture along which blocks of the Earth's crust have moved past each other—up, down, horizontally, or in some combination. Southern California has so many geologic faults that we tend to forget the crust can also deform by simple tilting, warping, or folding. Folds form best in well-layered, weak sedimentary rocks, of which Ventura County has a great thickness.

Bend a thick magazine up into an arch. It's easy to do because the pages slip past one another. If they were stapled together, the book would be harder to bend. It is the same with rocks. Well-layered sedimentary rocks fold as a result of tectonic deformation, whereas strong, massive rocks, such as granite, cannot be folded, so they fracture under stress, creating faults. Faults also form in severely stressed sedimentary rocks.

An arching fold is an anticline, so named because the opposing limbs incline away from each other, thus anti-inclined. To visualize an anticline, think of a rainbow with its ribbons of color—from deep red to violet. These individual colors represent layers of sediment—for example, clay, silt, sand, and gravel. An up-arching rainbow is akin to an anticline. Turn the rainbow over and you have a visual representation of a syncline, a trough-shaped fold, in which the opposing limbs incline toward each other, hence the term syncline, meaning together inclined. To remember which is which, "anticline" starts with A, which is shaped like an Anticline or Arch.

Oil, gas, and water can collect in anticlines if those fluids, along with porous beds to hold them, exist within the rocks. As these fluids rise to the crest of an anticline, they arrange themselves in order of density: gas on top, then oil, then water—all trapped beneath impermeable layers of rock. Petroleum geologists generally plan wells to tap into the oil zone rather than the gas. As much as possible, they preserve the gas below ground because its pressure helps force oil to the surface. Many anticlines are shaped like an inverted canoe with the central line, or axis, along the crest plunging down at both ends. Rocks with that structural form securely trap oil and gas. A few anticlines are cracked so they leak oil and gas; that is the cause of the natural oil pollution in the Santa Barbara Channel and the tar blobs along the shorelines near Goleta and Carpinteria.

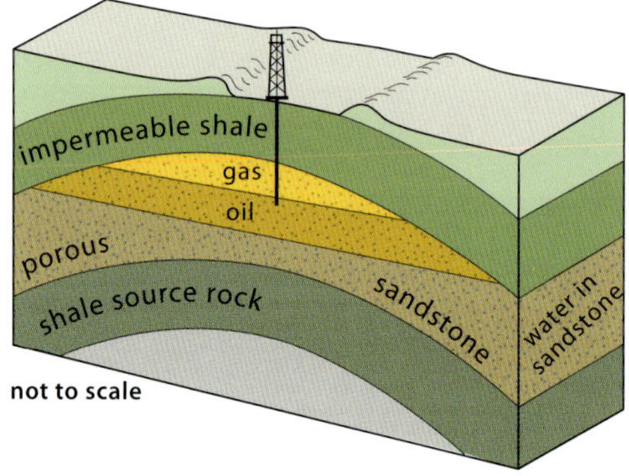

Gas and oil that rise into the top of a folded sandstone layer are trapped by an impermeable shale layer.

About 2.5 miles north of Ventura, along CA 33, a truly beautiful, large anticlinal structure extends east and west for about 15 miles. Dubbed the Ventura Avenue anticline, it is the site of a highly productive oil field, one of California's largest with a cumulative production of over 1 billion barrels of oil and several trillion cubic feet of gas. Most folds extend to great depths. Some geologists think the Ventura Avenue anticline ends on a horizontal fault at a depth of about 11,000 feet.

California geologists realized in the 1920s that the Ventura Avenue anticline had to be very young because many of the fossil seashells in the tilted beds are the same species as those still living along coastal beaches. They estimated the age of the fold as only 1 or 2 million years. Geologists in more stable parts of the United States regarded that figure as insanely young. Several kinds of evidence now show that even these early estimates of its age were far too old.

Obviously, a fold must be younger than the youngest beds it bends. Younger sedimentary rocks deposited in seawater just 200,000 years ago are also steeply tilted on the flanks of the Ventura Avenue anticline, so the fold must have formed since they were laid down. That is amazing! In most regions, oil-bearing folds are tens or hundreds of millions of years old. But stream terraces along the Ventura River,

Aerial view looking west in March 1981 at the road network, pumpjack pads, and landslide scars in the Ventura Avenue oil field.

which are only 16,000 years old, and which were also nearly flat lying when they were deposited, now dip gently where they cross the anticline, suggesting that the fold is still rising. Measurements described below confirm the active uplift.

The ages of the folded rocks and the amount of their folding imply that the Ventura Avenue anticline has risen at an average rate of nearly 0.6 inch per year for the last 200,000 years. A little simple arithmetic says that the fold crest has risen roughly 2 miles in that time. Rise of the fold crest has now slowed to less than 0.1 inch per year, but that is probably due to the changing shape of the fold, not to a slowing of the folding. As the limbs of a fold steepen, its crest rises more slowly. You can see how that works by laying a sheet of paper flat on a table, then pushing its edges together to make it into an anticline. At first, the fold rises rapidly, but as its limbs became increasingly vertical, its crest almost stops rising, even though you continue to push and bend the paper as fast as ever.

Level-line surveys show that the Ventura River valley rose over the anticline about 0.2 inch per year between 1920 and 1968. Between 1978 and 1991, the crest of the fold rose 0.25 to 1.5 inches relative to its flanks, or 0.1 to 0.2 inch per year. Some of that rise may be due to aseismic creep on faults within the anticline, but much of it is the result of folding. Furthermore, the ground above large oil fields typically *subsides* as oil, gas, and water are extracted. The absence of subsidence at the Ventura Avenue oil field indicates that the anticline is still rising, and no doubt about it: This is one of the few growing anticlines in captivity. Salute it!

If we can estimate that the Ventura Avenue anticline has risen about 2 miles in the last 200,000 years, then where is the 10,000-foot-high ridge that should have been created? Rocks on the surface of the Ventura Avenue anticline are so poorly consolidated and so easily eroded that you can dig into many of them with a shovel. Stream erosion has already stripped thousands of feet of rock as fast as the anticline has risen. If it were not so, the anticline would now make a ridge towering many thousands of feet above sea level.

On its way south to the coast, the Ventura River eroded a gap across the rising anticline, which CA 33 follows between Ventura and Ojai. The river maintains this course by eroding its bed as fast as the anticlinal ridge rises across its path. Raised remnants of old floodplains of the Ventura River survive as terraces, broad steps above the modern floodplain extending to the valley walls. When the river was flowing on them, these old valley floors sloped the same direction as

View looking northwest across the Ventura River valley from Grant Park. The green agricultural patches are on remnants of stream terraces at two levels on the west bank.

the river flows, gently downstream toward the sea. The rising fold has tilted the terraces on its north limb so they now slope upstream.

Marine terraces on the south limb of the anticline are old wave-cut benches that originally sloped about 1 degree seaward, the usual slope of such terraces. They are two to three times older than the stream terraces on the north flank of the fold. The marine terraces now slope about 10 degrees seaward and stand hundreds of feet above sea level. Part of that rise is due to the folding, the rest to displacement along faults.

Folded sedimentary beds in the Ventura Avenue anticline contain several thin layers of volcanic ash, whose age is generally possible to determine through chemical analyses and microscopic studies to match with volcanic rocks elsewhere. Then, it becomes possible to obtain isotopic ages of those volcanic rocks associated with the ash, thus fixing its age.

A layer of Bishop Tuff, erupted from the Long Valley caldera near Mammoth Lakes 765,000 years ago, is interbedded with steeply dipping marine sedimentary beds in the Ventura Avenue anticline. Both the tuff and those beds were flat lying at the time of their deposition and have been subsequently tilted. The well-known Lava Creek ash layer from the Yellowstone caldera lies within steeply tilted beds on the south limb of the anticline. It exactly matches ash erupted during a gigantic volcanic explosion from the Yellowstone area 631,000 years ago. About 3,000 feet of younger sedimentary rocks above that ash are also tilted. Other less exact dating techniques and the estimated rate at which sediments are accumulating on the seafloor off California suggest that the anticline began to grow in just the last 200,000 years, an astonishingly short time for such a major geologic structure.

This is all very interesting in an abstract way, but can any of the anticline be seen? Certainly, although erosion has erased most of it. Drive north from Ventura on CA 33. When passing the first exit, Stanley Avenue, about 1.5 miles north of US 101, peek here and there through the trees and see a large scar on the east valley wall. There, soft shale beds have been excavated to make drilling mud for the oil field and to provide raw material for an expandable clay plant. The sedimentary layers in the face of the scar, well out on the south limb of the anticline, dip to the south at an angle of about 30 degrees.

A little farther north, a high levee blocks your view of massive sandstone beds in the west valley wall that are less steeply inclined, also to the south. Near the Shell Road exit, sandstone layers dip even more gently down to the south on both sides of the road, although trees block most of the view. Farther ahead, layers exposed in the cuts on the east valley wall, opposite the large sign announcing "Cañada Larga Exit, 1.5 miles," are nearly horizontal.

Horizontal makes sense at that place because beds at the crest of an anticline are flat lying. The trend of this fold's central line is slightly oblique to the Ventura River and CA 33, so the corresponding horizontal beds lie somewhat farther south on the west valley wall. The first beds you see west of the Cañada Larga sign dip gently to the *north*. That means we have crossed the crest of the fold.

Exit at Cañada Larga Road. Stop at the bottom of the off-ramp, and then turn west (left) under the overpass to the next boulevard stop. Turn south (left) and go south on what becomes Ventura Avenue. You pass the north end of Crooked Palm Street on the west (right) 100 feet before crossing under the freeway; the south end of Crooked Palm comes up in another 0.9 mile. Drive across the crest of the anticline where Ventura Avenue passes all the oil pumpjacks at Shell Road.

As you continue south on Ventura Avenue, look east beyond Shell Road to see some exposures of massive sandstone beds dipping south at an angle of 20 to 30 degrees. You are now well out onto the south limb of the fold. Between Seneca Street and Stanley Avenue, you again see the big cut on the east valley wall, with beds dipping steeply to the south.

Between 1895 and 1898, Ralph B. Lloyd, a Ventura resident and recent graduate from the University of California at Berkeley, did essentially what we have just done but on a horse. He knew what anticlines looked like and was aware that oil and gas accumulate in such structures. Ralph Lloyd became a dogged advocate of drilling for oil on the Ventura Avenue anticline. His was a long, rocky road, but finally, on September 25, 1916, the well designated Lloyd No. 2 came in with the first significant oil production. Shallow wells had earlier (1903) produced gas. Lloyd later drilled fifty-seven successful wells before his first dry hole, a record any operator could envy. Ralph Lloyd became the father of the Ventura Avenue oil field because he recognized the significance of those tilted beds—and he also became a very wealthy man.

Regional Vista at Grant Park

Turn left off Ventura Avenue at Ramona Avenue, 0.7 mile south of Stanley, and head for Cedar Street at the base of the hill. Follow Cedar Street to the south (right) and uphill. Near the top, about 100 feet before a center divider splits Cedar Street, turn left (north) onto Ferro Drive and follow it to Grant Park and the Padre Serra cross by keeping left at all intersections. If you miss Ferro Drive, continue on Cedar Street, which shortly becomes Poli Street. In a block or so, at the large county courthouse and jail buildings, Brakey Road turns narrowly and very steeply to the north (left). Follow it under a large stone arch to an intersection with a sign that directs travelers west (left) to Padre Serra cross and Grant Park, a superb spot (34°17.076N, 119°17.773W) for viewing the geologic features around Ventura. Walk to the northwest corner of the parking area and descend to the dirt walkway outside the upper hedge and its retaining wall. Follow the walkway counterclockwise around the parking area.

Look directly up the Ventura River valley and think of that direction as 12 o'clock on a horizontal clock face. Across the Ventura River at 10:30, you see remnants of two stream terraces, and at 9 o'clock the smooth skyline profile of the marine terrace that now dips 10 degrees seaward.

The Ventura Avenue anticline, especially its eastern part, may be seen partway around the path between 10 and 11 o'clock. Since drilling rigs no longer remain over wells, as in earlier years, identification of the oil field and anticline is not easy. The crest of the anticline is beneath a wide belt of brush and scattered oak trees that contrasts with the mostly grassy slopes of the limbs on either side. The broad, flat floor of the Ventura River narrows to a gap through massive sandstone beds in the core of the anticline. The west half of the oil field at 9 o'clock has many secondary hillside roads, cut banks at well sites, tanks, and large gas compressor plants toward the east end. Not all the hillside scars are artificial. Some are the scarps at the heads of landslides, which have given oil companies fits by shearing off producing wells 100 to 200 feet below the surface.

Farther around the path and looking south, you can see a pond at the mouth of the Ventura River, Ventura harbor, its pier, and the beach. Offshore to the west and south is Santa Barbara Channel, and on the far skyline two of the seven Channel Islands: Anacapa on the east and Santa Cruz on the west. At 4 o'clock beyond the steps to Padre Serra cross is a good view of the lovely curving strand of Pierpont Bay. The bay view gives way south-southeastward to the long strand extending past the breakwater at Ventura Marina. You may need binoculars to see the decommissioned Mandalay Beach power plant with its lofty stack at 4 o'clock and beyond to Point Mugu at the south tip of the strand.

View looking south-southeast from Grant Park. The Ventura pier extends into Pierpont Bay, and the dark structures on the bay's curving beach are groins to impound southeast-moving sand. Oxnard Plain and the west end of the Santa Monica Mountains are on the horizon.

PIERPONT BAY

From 1855 until at least 1933, the beach at Pierpont Bay was building outward into the ocean at an average rate of about 7.5 feet per year. After the Ventura River was diverted for irrigation and flood control, it could no longer deliver much sand to the coast. That deprived the Pierpont shoreline of its steady supply of sand, and the waves then eroded the beach some 200 feet inland within fifteen years. Twenty-two acres of beach were lost at a value of $50,000 per acre—that's over $11.5 million in 2020.

Look along the coast for groins, rock walls that project from the shore into the surf. They are intended to impound sand moving southeast along the beach and to prevent the loss of more beach. They do those things to some degree, yet the gains they impart become someone else's loss. The sand impounded by the groins doesn't move farther down the coast, so the structures deprive beaches farther along the shore of their share of sand.

A healthy stretch of beach is in a state of dynamic equilibrium between the sand that comes in one end and the sand that moves out the other. They balance. That is why a beach can appear unchanged even though sand flows along it as though it were water flowing down a river. Groins, because they disrupt the natural flow of sand, are not perfect solutions to the beach deterioration that follows river diversion.

SANTA MONICA MOUNTAINS AND THE CHANNEL ISLANDS

Look farther southeast to see Point Mugu and the west end of the Santa Monica Mountains rising abruptly in skyline silhouette above the Oxnard Plain. The relatively smooth skyline extending east from Point Mugu continues a considerable distance before it abruptly rises into the ragged skyline of Old Boney Mountain, a brooding mass that frowns down on the Oxnard Plain to the west. Sedimentary beds underlie the lower, smoother skyline, whereas Old Boney is held up at a higher and more ragged skyline by harder and more resistant volcanic rocks, the Conejo Volcanics, which underlie much of the subsurface of the Oxnard Plain and correspond to volcanic rocks on Anacapa and Santa Cruz Islands.

The Channel Islands, the westward extension of the Santa Monica Mountains, are part of the east-west trending Transverse Ranges, now rotated about 100 degrees clockwise from their original position adjacent to the mainland coast between San Diego and San Clemente. Major faults cutting the islands are extensions of mainland faults of the Transverse Ranges. Some of the oil-bearing structures in the Santa Barbara Channel—Rincon anticline, for example—are the extensions of mainland structures, including the Ventura Avenue anticline.

If the air is clear down to water level, you may be able to see the famous sea arch at the east end of Anacapa Island. The arch is actually in a large offshore rock, but from this view, it appears to be part of the island.

Look from Point Mugu west along the length of the chain of the four Channel Islands. From east to west, they are Anacapa, the smallest; Santa Cruz, the largest; Santa Rosa; and San Miguel, which is too far to be seen from this vantage point. It is easy to imagine that a drop in sea level would convert all of them into a long peninsula projecting west from the Santa Monica Mountains, converting Santa Barbara Channel into a bay. This hypothetical peninsula has been called Santarosae and also the Cabrillo Peninsula to honor Juan Rodríguez Cabrillo, the Portuguese-born, Spanish soldier and explorer who sailed the California coast in 1542 to 1543.

Cabrillo Peninsula probably did exist during the last ice age, when sea level was some 300 feet lower than it is now. It became the present string of islands as the melting glaciers returned their water to the ocean about 12,000 years ago. The ice age left some surprising souvenirs: Fossil bones of pygmy mammoths (*Mammuthus exilis*) have been discovered on the islands. A replica of the fossils is displayed in the visitor center at Channel Islands National Park headquarters in Ventura. Not a trace of similar pygmy mammoths is known on the mainland. Some investigators suggest that big Columbian mammoths swam the channel to reach the islands—because modern elephants are good swimmers—then evolved there into a race of pygmies. Distances from the mainland to the islands and among the islands would have been shorter when sea level was lower than today, and it is easy to imagine them walking along the peninsula—or swimming short distances—with something to eat on the way. Then, when sea level rose and converted the peninsula into a chain of islands, they found themselves isolated. Why did they become pygmies? Probably because the limited food supply of the islands could not support big animals.

The small grove of rare Torrey pines (*Pinus torreyana*) on Santa Rosa Island may be another such memento of the Cabrillo Peninsula. The trees could have migrated much more easily along a peninsula than across a 20- to 30-mile-wide sea strait. The trees, relics of the ice ages, are rare in today's warmer and drier climate.

A replica of the pygmy mammoth skeleton found on Santa Rosa Island in 1994 is on display at Channel Islands National Park visitor center in Ventura.

VIGNETTE 5

SANTA BARBARA HARBOR

Perils of Tampering with Nature

SANTA BARBARA COUNTY

California's rugged coast has few good harbors, mainly because it is rising; rising coasts tend to have straight shorelines without protected indentations. Artificial anchorages, such as the one at Santa Barbara, are harbors only in name, a form of political license.

The Santa Barbara area has long been favored by people of comfortable means. Some owned yachts in the early days; many more own small sailboats today. Commercial enterprises include a sizable fishing fleet and excursion boats for travel along the coast and to the Channel Islands.

Interest in an anchorage here dates back to at least 1850, but little was accomplished until 1926 when Major Max Fleischmann, of yeast fame, wanted an anchorage for his yacht closer to his home in Santa Barbara than in Goleta Slough, 10 miles to the west, which seemed to offer the only possible site for a local harbor. Fleischmann offered a sum of $200,000, to be matched by public funds, to construct a breakwater about 600 feet offshore, parallel to the beach. The site selected was off Point Castillo, which was referred to as the Castle Rock site because of a ragged offshore rock that was later demolished. The breakwater lies about 1,500 feet southwest of Stearns Wharf, a structure erected in the 1870s and subsequently modified several times. Blocks of volcanic rock were hauled by barge from Santa Cruz Island to build the breakwater.

The coast at Santa Barbara trends nearly due east and west, and large volumes of sand are carried eastward along this shore by waves and longshore currents. Some people realized this from the beginning. The longshore sand flow occurs because most of the impinging swells are generated by storms far out at sea to the northwest. Although bent by refraction as they move into shallow water, these swells remain slightly oblique to the beach in an easterly direction. This angled wave attack creates an eastward longshore drift. Occasional storm waves from the southwest temporarily reverse the normal longshore drift direction.

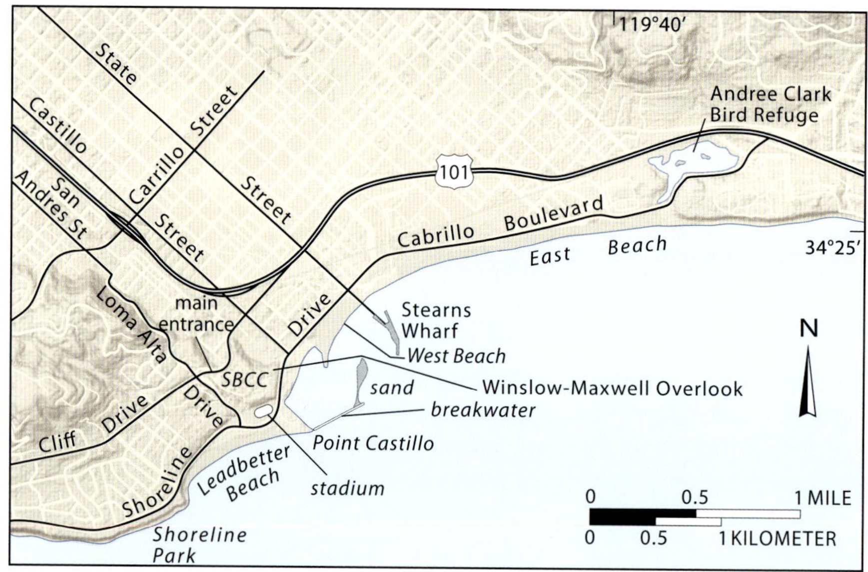

SBCC stands for Santa Barbara City College.

GETTING THERE

Santa Barbara Harbor and its setting can be viewed nicely from the campus of Santa Barbara City College. The best time is weekends when school is not in session. Enter the campus through the kiosk on the south side of Cliff Drive. If entering Santa Barbara from the east via US 101, turn off at Cabrillo Boulevard (exit 94-C), a left-hand exit, and proceed 3 miles west on Cabrillo past Stearns Wharf and the boat harbor to Loma Alta Drive. Turn northwest (right) there and at the top of the hill turn east (right) at Cliff Drive. In a hundred feet or so, turn south (right) at the entrance kiosk of the City College campus.

From the west on US 101, exit at Castillo Street (exit 97), proceed southeast (right) on Castillo for four blocks past the baseball fields and Pershing Park to the traffic light, turn southwest (right) onto Shoreline Drive, and continue about a half mile to the traffic light at Loma Alta Drive. Turn northwest (right), proceed to the top of the hill, turn east (right) onto Cliff Drive, and then make a sharp hairpin turn right to the entrance kiosk.

If the kiosk is occupied, explain to the attendant that you are studying geologic relationships at Santa Barbara Harbor that are best viewed from the Winslow-Maxwell Overlook at the east tip of campus. If the kiosk is unoccupied, proceed slowly straight ahead across speed bumps, past the MacDougall Administration Center, the Occupational Education building, and the Earth and Biological Sciences building until the campus opens out and the road curves northeastward past the Bookstore and toward the mountains. Continue past several low, shed-like buildings on the right until you reach the International Education Center on the left. Park near the turn-around at the hillock with its four palm trees and the Winslow-Maxwell Overlook (34°24.424N, 119°41.698W), for a splendid view north over Santa Barbara and the Santa Ynez Mountains.

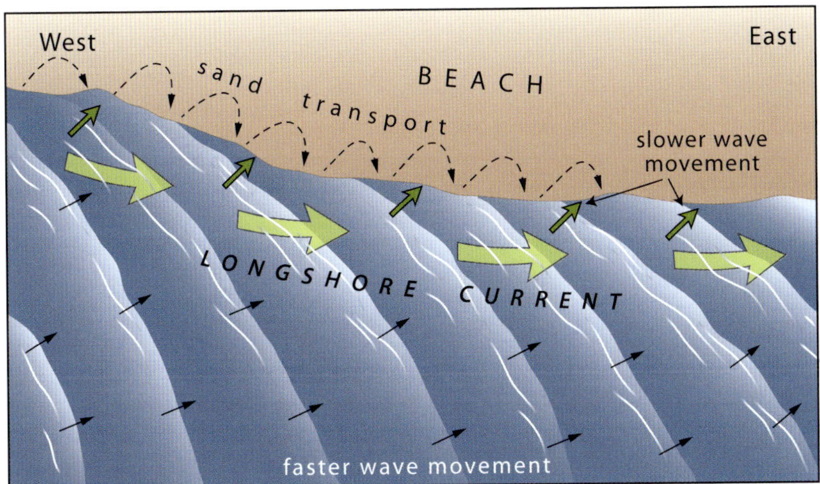

Waves approach the shore and carry sand up the beach at an oblique angle, but both return to the sea perpendicular to the coast as backwash. Sand grains move along arc-shaped paths. The net effect is eastward longshore transport of both water and sand.

Construction of the offshore breakwater was complete by 1929, and photographs show that sand soon began to accumulate in the harbor. The original expectation had probably been that an offshore breakwater would permit longshore drift to carry sand around the seaward side of the breakwater, leaving the harbor unobstructed. Unfortunately, principles of ocean-wave refraction were not fully understood or appreciated in the 1920s. The west end of the breakwater interfered with the longshore drift, so that swells were bent westward, the harbor basin filled with sand, and a broad sandy beach—Leadbetter Beach—formed from the shore into the west end of the anchorage area.

To rectify the situation, Major Fleischmann gave an additional $250,000 to construct an elbow that attached the west end of the breakwater to the beach at the foot of Point Castillo. This L-shaped structure, completed in 1930, successfully protected the harbor from further sand deposition from the west—for about three years.

The primary results were, of course, continued deposition of a huge quantity of sand that widened Leadbetter Beach, west of Point Castillo, and formed West Beach. The broad sand flat, now occupied by extensive parking areas and La Playa sports stadium, was deposited in the remarkably short period of less than ten years.

Aerial view looking northeast of Santa Barbara Harbor elbow under construction, October 31, 1929. Stearns Wharf lies beyond the breakwater. The white sand projection in the harbor formed by waves refracting around Point Castillo (white arrow). Note that waves lapped against the sea cliffs that formed the shoreline at this time. —Fairchild air photo O-139, Department of Geography, University of California, Los Angeles

Aerial view of Santa Barbara Harbor looking north on December 17, 1931, two years after completion of the breakwater. Note the accumulation of sand west of Point Castillo (black arrow), forming Leadbetter Beach. —Fairchild air photo O-2700, Department of Geography, University of California, Los Angeles

Aerial view of Santa Barbara Harbor looking northwest on November 27, 1934, five years after completion. Sand has filled in Leadbetter Beach seaward of Point Castillo (black arrow). —Spence air photo E-5780, Department of Geography, University of California, Los Angeles

Aerial view of Santa Barbara Harbor looking west-northwest on April 8, 1964. The sports stadium west of Point Castillo (black arrow) was constructed on beach sand at the base of the former sea cliff. West Beach has formed between Stearns Wharf and the harbor. —Spence air photo E-18971, Department of Geography, University of California, Los Angeles

After the area west of the breakwater elbow filled with sand, longshore drift built a shallow sandbar along the outer side of the breakwater. By 1940, or somewhat earlier, the bar extended to the breakwater's northeast end. There, refracted swells happily seized the drifting sand and dumped it into the harbor, creating shallows and eventually a bar across the harbor entrance. Major storms from the southwest sent their waves to join the process of shoveling sand into the harbor in spite of a new 1,000-foot-long northeastward extension of the breakwater toward Stearns Wharf.

Meanwhile, East Beach and other prized beaches farther east were starved for sand. They deteriorated rapidly as hungry waves chewed at them, narrowing them and exposing rocks where before there had only been sand. Eventually, this deterioration extended to a critical degree as far east as the fine beach at Carpinteria, much to the displeasure of the local populace. According to one report, beaches as far as 45 miles down the coast also deteriorated because of the Santa Barbara Harbor perturbation. Hotels in Montecito, adjacent to formerly sandy beaches, brought suit against Santa Barbara, which also suffered as waves threatened to undercut Cabrillo Boulevard and a newly constructed pavilion on East Beach.

So, what to do? One obvious action was to clean out the harbor and remove the sandbar blocking its mouth by dredging. Intermittent dredging began in 1935 and continued irregularly through the 1950s. By the early 1960s, a year-round dredging program was initiated that continues to this day with federal funding. The dredged sand is carried in a pipe as a water-rich slurry and dumped into the surf east of Stearns Wharf for redistribution eastward. East Beach has been restored, and the beach at Carpinteria has partly recovered. Beaches farther east along the Rincon are still bereft of sand. To see a sand-starved, wave-battered, ravished waterfront, visit the eastern part of the Solimar settlement along the Pacific Coast Highway on the Rincon between Ventura and Santa Barbara.

Santa Barbara Harbor and other coastal constructions are not the only, or even the major, factors causing the deterioration of southern California beaches. Flood-control dams and debris basins on coastal streams trap sediment and prevent it from reaching the ocean. For example, dams on the Santa Ynez River have reduced the sand supply from this river by 68 percent. The practice of damming rivers has been to restrict, if not eliminate, floods because they are dangerous, destructive, and wasteful, particularly in urbanized areas. We do so, however, at the price of serious deterioration of our sandy beaches. The

A dredge removes sand that drifts around the northeast end of the breakwater into the harbor entrance between the breakwater and Stearns Wharf and transfers it by an underwater pipe to the other side of the wharf. (34°24.44N, 119°41.33W)

problem of beach deterioration is currently being attacked by small-scale improvisations. The long-range solution lies in providing greater supplies of sand to the waves devouring the beaches, but satisfactory sources for that sand remain a problem. Recent interest has focused on the possibility of dredging sand from offshore accumulations and dumping it into the surf for natural transport along the shore.

Santa Barbara Harbor is currently in a state of quasi-equilibrium thanks to continuous dredging, an expensive process. In problems like this, we need to work with nature rather than against it. Model studies in experimental tanks could determine what, if any, configuration and arrangement of breakwater structures could provide a protected anchorage and still allow free passage of sand by longshore drift. It is not clear that such a configuration can be designed, considering the many complex variables involved, but study of the possibility is a challenge worth addressing because many other harbors face similar problems.

For close views of the harbor, head to the Winslow-Maxwell Overlook on the campus of Santa Barbara City College (See Getting

There). A paved and dirt pathway will take you west about 500 feet through the Lifescape Garden and Chumash Ethnobotanical Preserve and along the edge of the bluff to the north edge of the sports stadium. Along the way are views here and there of the harbor. The bluff is a sea cliff whose feet were bathed by waves before completion of the full harbor breakwater in 1929.

The story of Santa Barbara Harbor can be worked out by studying the photos and digesting their captions. Remember that everything you see transpired since 1930, and most of it before 1940. It is sobering to realize that the broad flat extending out to Leadbetter Beach, upon which the sports stadium sits, built up in just a decade solely by the accumulation of sand carried by the longshore drift.

Few people deny that the Santa Barbara anchorage is of major commercial and recreational value to the area. The harbor requires much supervision and maintenance and is vulnerable to devastation by storms from the southwest. The search continues for the best solution to operate this valued possession at reasonable cost with minimum impact on the natural environment.

Aerial view looking northeastward across Santa Barbara Harbor in 2018, eighty-nine years after completion of the breakwater in 1929. All of the flat ground in this photo, including the parking area, football field, boulevard, and beach, formed by deposition of longshore sand drift in one decade (1930–1940).

Natural systems are interdependent, and jiggling one factor may cause others to shift in unpredictable ways. When humans tamper with natural systems, they often initiate chain reactions that produce undesirable and unexpected results. That lesson about disruption of a natural system and attempts to manage an altered system is very clear in Santa Barbara Harbor. We must proceed gently, carefully, and with as much knowledge and understanding as possible when we tamper with dynamic natural systems, longshore sand drift being just one example.

A walk out on the breakwater is fun, so do it. Go back to Loma Alta Drive, descend to the base of the bluff, and turn north (left) on Shoreline Drive. Turn east (right) onto Harbor Way and proceed straight ahead. Except on weekdays, parking is usually available on either side among the buildings ahead. Follow your nose and sense of direction to the walkway leading onto the breakwater.

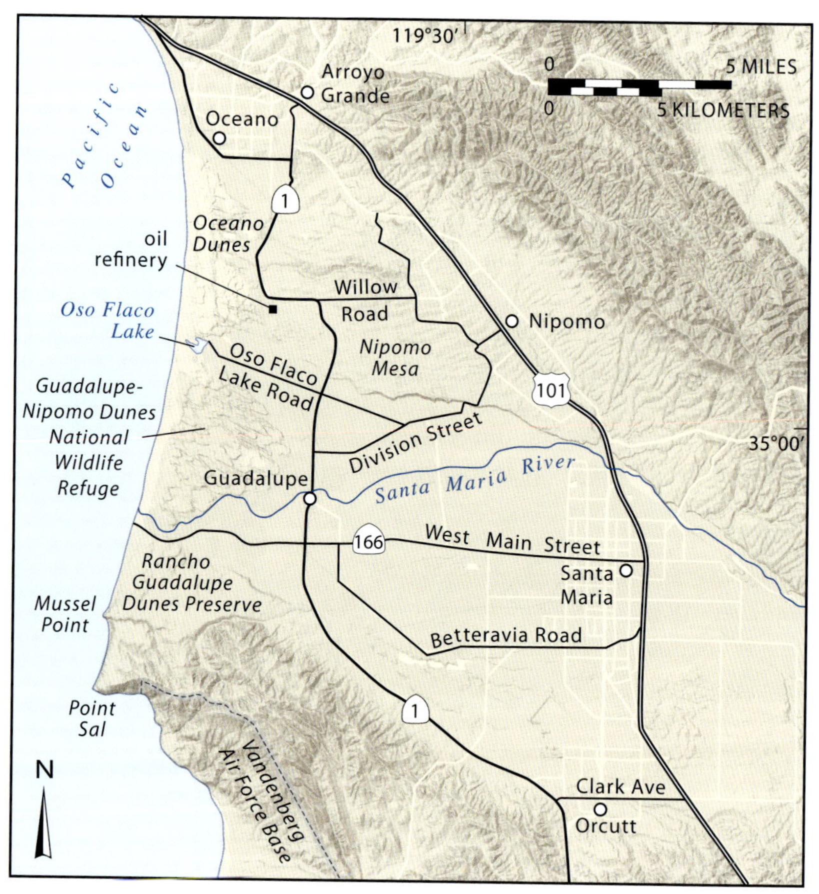

GETTING THERE

Many good primary and secondary roads provide access to the 35 square miles of the Guadalupe-Nipomo dune fields. The most accessible dunes are in Rancho Guadalupe Dunes Preserve, a Santa Barbara County Park. To reach them from US 101 in Santa Maria, take CA 166 (West Main Street, exit 171) and drive west on West Main Street about 9 miles to CA 1. Continue west on West Main Street for another 5 miles to the 600-acre park at the end of the road. Access to the Rancho Guadalupe dunes is seasonal, limited to the months of October through February, whereas the access road and beach are open year-round. Dogs, pack animals, and drones are prohibited.

VIGNETTE 6

GUADALUPE DUNES

An Ice Age Sand Lobe

SANTA BARBARA AND SAN LUIS OBISPO COUNTIES

The Guadalupe-Nipomo Dunes National Wildlife Refuge, home to more than 120 species of rare plants and animals, contains some of the most remote and least disturbed habitats in the dunes complex. Oceano Dunes State Vehicular Recreation Area, north of and adjacent to the Guadalupe-Nipomo Dunes Complex, is the only area in California where motor vehicles may drive right on the beach and into the adjacent dunes. The Oso Flaco Lake Natural Area is nearby.

Mention sand dunes and many people think of deserts, while others have visions of seashores. Both are correct; dunes favor these two environments. Large masses of sand accumulate into dunes only where there is a constantly renewed source of sand and a strong prevailing wind that blows long enough (many days per year) and hard enough (at least 25 miles per hour) to move sand freely. Ocean beaches without vegetation to break the wind are ideal sources of sand, with waves constantly replenishing the supply.

A strong prevailing wind blows across most coasts, typically from sea to land. It carries sand inland, but usually only to where vegetation breaks its force, inhibiting sand transport. Wind moves sand off any beach, even a wet one; wind dries grains on the surface so they no longer stick together. Watch the next time you see a stiff breeze sweeping across a wet beach. Removal of dry, loose sand from any source is a breeze for wind.

Coastal dunes differ from most desert dunes in the large role plant cover plays in determining their location, size, configuration, and eventual stabilization. Plants rarely stabilize any large desert dune, but they almost invariably tie down coastal dunes before they are very old.

The California coast boasts at least twenty-seven dune fields of respectable size. Some are narrow belts bordering the inland edge of the beach; others form broad sheets that extend 1 or 2 miles inland. A few, such as the large dune field that borders Monterey Bay south

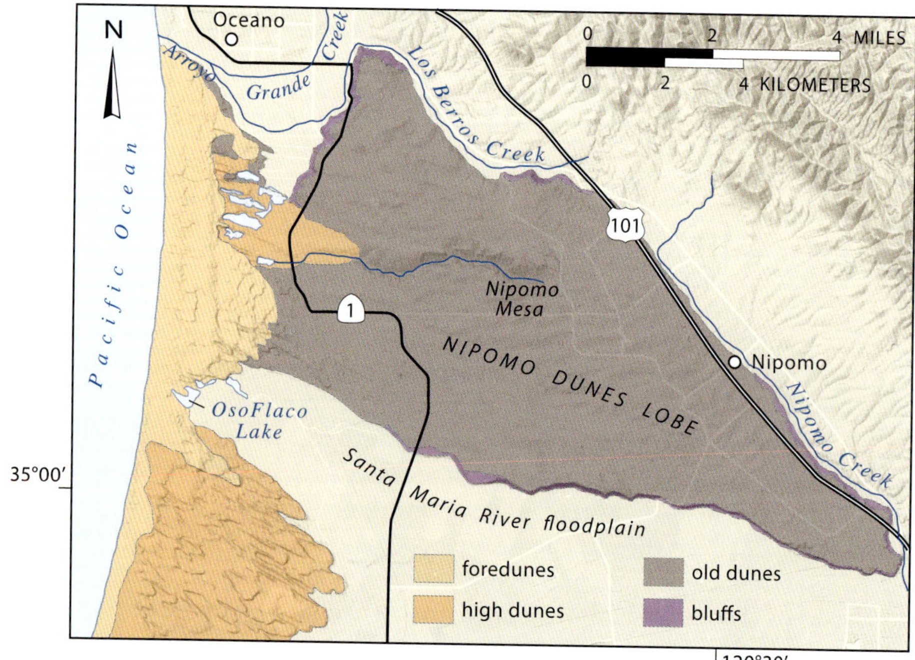

Map of Nipomo Dunes lobe and relationships among dune types.
—Modified from Cooper, 1967

of the Salinas River in central California, cover more than 100 square miles and reach 5 to 6 miles inland.

The broad lowland where the Santa Maria River meets the sea is richly endowed with dunes. Three large tongues of dune sand extend many miles inland. The Guadalupe-Nipomo Dunes Complex is the second largest remaining dune system in California and one of the most spectacular lobes of dune sand along the entire coast of Oregon and California. Stretching more than 18 miles along the southern San Luis Obispo County and northern Santa Barbara County coastline, the dunes extend fully 12.5 miles inland to a little east of US 101 at Nipomo and Santa Maria.

Why are dunes so abundant on this part of the California coast? The answer involves an interplay of sand supply, prevailing wind direction, and offshore ocean currents. The Santa Maria River brings sand and rock fragments to the coast, where the constantly sloshing surf grinds them into particles small enough for the prevailing northwest winds to blow onshore. Longshore currents flow downcoast and around Point San Luis only to reverse direction nearshore as they

Aerial view looking southwest at the Guadalupe-Nipomo Dunes Complex between Point Sal (upper left corner) and Oceano. Active foredunes (white) encroach upon an ancient stabilized Nipomo Dunes lobe in the Santa Maria River floodplain. —Photograph by Doc Searls

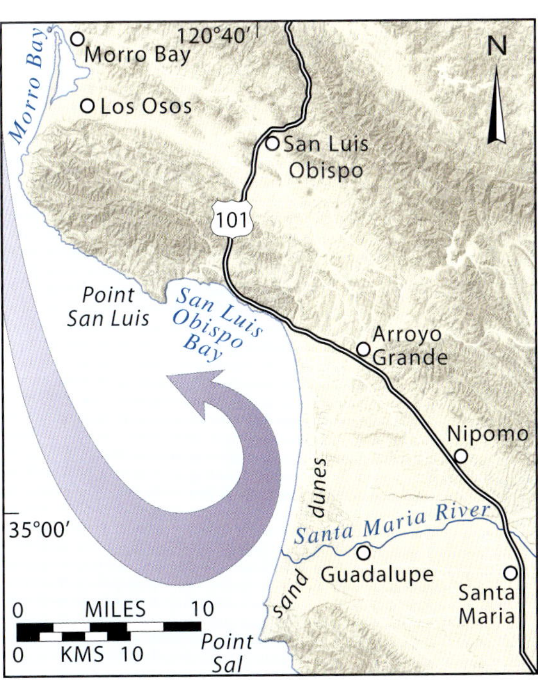

Sand, delivered to the coast by the Santa Maria River, is directed upcoast by a counterclockwise eddy in the generally south-flowing longshore currents in San Luis Obispo Bay. The sand piles up in the bay and is redistributed along the shore by waves and wind.

Vegetated sand mounds in the active foredune belt commonly form above and around obstacles, such as logs or boulders, which obstruct sand transport. Low-growing vegetation catches hold and traps the sand. The plants must continue growing upward to keep from being drowned by sand, thus trapping ever more sand, so that the mound also grows. (34°57.81N, 120°38.98W)

eddy in San Luis Obispo Bay. Thus, instead of the sand being carried downcoast from the river's mouth, ocean currents continually circulate it back into position to be blown ashore.

The Guadalupe-Nipomo Dunes Complex can be separated into at least three distinct parts that differ in appearance and, presumably, age. The youngest belt, called foredunes, comprises active dunes with little or no vegetation growing on them. They are light colored, as seen from the air, and extend no more than 1.5 miles inland from the beach. Their dune forms are sharp and fresh, constantly changing.

The foredunes come to an abrupt and steep inland edge where they encroach irregularly upon the older, more subdued high dunes, most of which lie under a cover of vegetation. The advancing foredunes surround hollows among the older high dunes to create dozens of depressions that hold shallow ponds and marshes. Fresh groundwater keeps them full. It is unusual but refreshing to find such wet ground in a dune field. Bird watchers find rich and varied bird life along the margins of these waters.

Chaparral vegetation on inactive, high dunes. (35°01.84'N, 120°37.46'W)

The steep front of an active foredune encroaches (from right to left toward the road) onto a vegetated hollow. (34°57.70N, 120°38.91W)

Sand dunes advance when sand is blown up the gentle windward slope of an active dune and then avalanches down the steep lee slope (Vignette 17). The inclination of lee slopes is about 34 degrees—the angle of repose—which is the maximum angle of a slope at which loose, cohesionless material will rest on a pile of similar material.

Sand avalanche scars on the steep lee slope of a foredune. The tire track is about 4 feet long. (34°57.72N, 120°38.9W)

A separate mass of linear, somewhat older dunes, the high dunes, extends about 2.5 miles farther inland than the active foredunes along the coast. Their sparse plant cover is enough to stabilize them. These dunes consist of long, nearly parallel sand ridges that may join at their downwind ends. The wind blows sand out of the troughs and onto the adjacent ridges where vegetation stabilizes their margins. Linear dunes in the Nipomo area are exceptionally long and sharply pointed, aligned with the prevailing strong west and northwest winds.

The oldest and largest part of the Guadalupe-Nipomo Dunes extends east and southeast from the active foredunes along the coast and the stabilized sand sheet. These forms hardly resemble dunes but instead are a disorganized assemblage of rounded hillocks and hollows. Here and there where CA 1 proceeds across Nipomo Mesa are exposures of homogeneous sand with the uniform grain size characteristic of dunes. At most, the hills stand about 80 feet above the hollows. The degraded shapes of these dunes suggest that they have been covered by vegetation for many years. Iron oxide cements the dune surfaces into

Linear sand dunes in the Nipomo area are oriented parallel to the prevailing northwest onshore wind direction and highlighted by the vegetation on their crests. —Google Earth image

a crust in places and stains the sand a dark brown. The plant cover on Nipomo Mesa is mostly scrubby chaparral brush typical of so much of southern California, with a few struggling oak trees. In the Santa Maria area, the oldest dunes are covered with housing developments west of US 101 between Clark Avenue and Betteravia Road and are heavily modified by agricultural activities east of US 101.

UCLA geographer Anthony Orme pointed out that the ancient and modern dunes are similar in orientation, structure, and mean sand grain size, indicating that similar dune-forming conditions have persisted along this coast over the past several hundred thousand years.

Much of the oldest sand sheet on Nipomo Mesa is now urbanized. Roads, homes, ranches, excavations, lemon and avocado groves, and the replacement of the original vegetation with imported trees, including large plantations of eucalyptus, have drastically modified the original surface. But 2,553 acres in the heart of the old, 22,000-acre Nipomo sand sheet are now a US National Natural Landmark and a US National Wildlife Refuge, protected and managed since 2000 through the efforts of several public and private agencies, including the San Luis Obispo and Santa Lucia Chapters of the Nature Conservancy and the US Fish and Wildlife Service.

The Santa Maria River trimmed the south margin of the Nipomo Dunes to form a steep cliff. Converging furrows on the floodplain in the foreground are for agricultural purposes. (35°00.635N, 120°30.855W)

We've now established that the Guadalupe-Nipomo Dunes consist of three different episodes of dune formation separated by significant time intervals. When we tell you that this may reflect the behavior of glaciers during the last ice age, your reaction is likely to be, "Well, the old boys blew it this time." Let us defend ourselves.

If all the world's present glaciers were to melt, sea level would rise more than 200 feet. When the enormous ice age glaciers formed at various times during the last million years, they tied up enough water to drop sea level as much as 450 feet, depending upon which ice age was involved. The modern dunes probably began to form a few miles seaward of where they are now during the Last Glacial Maximum, about 20,000 years ago, when sea level was several hundred feet lower than today. Then, as the glaciers melted and sea level rose, waves attacked and reworked those dunes, and winds blew the sand inland to form dunes in their present location.

Older dune masses now well above present sea level, such as those on Nipomo Mesa and in the Santa Maria area, are thought to be relics of even higher stands of sea level between ice ages. To make the much larger older part of the dune sheet would require a long interval of high sea level, perhaps more sand than the modern rivers and beach supply, and possibly stronger winds than those that now blow. What we know of the record of the great ice ages and the intervals between them suggests that such conditions may have existed about 125,000 years ago when sea level was about 20 feet higher than present. Perhaps that was when the wind blew the oldest part of the Guadalupe-Nipomo Dunes inland from the beach.

Exploring the Dunes

Start your excursion with a visit to the Guadalupe-Nipomo Dunes Center, an agency that promotes the conservation of the dunes' ecosystem through education, research, and cooperative stewardship. It is located at 1065 Guadalupe Street, at the north end of the city of Guadalupe, and features exhibits about the natural history of the dunes and the area's cultural history.

With the center's map in hand, proceed 1 mile south to West Main Street (CA 166), turn right, and continue 3 miles to the Rancho Guadalupe Dunes Preserve gate, then follow a 2-mile-long access road through extensive sand dunes to a parking area at the beach. There, you will have dynamic views of the south end of the dune complex and the Santa Maria River estuary as well as access to miles of broad, sandy beach. Walk inland out onto the dunes and study the various surface features, including ripples and the multitude of tracks and trails of diverse animals.

Oso Flaco Lake Natural Area, a hidden gem in the California State Park system, is part of Oceano Dunes State Vehicular Recreation Area. To get there from CA 1, drive 3 miles west on Oso Flaco Lake Road, 4 miles north of Guadalupe. From the fee parking area at the end of the road, it's an easy 0.25-mile walk along a paved road through a

THE MANY USES OF SAND

You will pass a sand quarry on the access road to Rancho Guadalupe Dunes Preserve. Its sand is used for playgrounds, golf courses, sandblasting, and many industrial applications, including filters, abrasives, ceramics, absorbents, and the manufacture of concrete. Construction and industrial grade sand and gravel account for nearly one-third of the value of California's nonfuel natural resources.

Aerial view, looking north, across Mussel Point headland, sand dunes, and coastal strip of Rancho Guadalupe Dunes Preserve. The mountains on the skyline enclose San Luis Obispo Bay. —Photograph by Cameron Venti on Unsplash

Aerial view looking west across Oso Flaco Lake, one of the water bodies dammed by the light-tan area of advancing foredunes between the ocean and the older, vegetated high dunes.

thickly wooded area. The road emerges from the trees at Oso Flaco ("skinny bear") Lake. A boardwalk crosses the lake and then enters the zone of stabilized, high dunes. Continue along the boardwalk to a high viewpoint to gain splendid views of the foredunes along the long remote beach and surrounding landscape.

CITY OF THE PHARAOH

In 1923, pioneer filmmaker Cecil B. DeMille built the largest and most lavish set in movie history in the Guadalupe-Nipomo dunes for his silent epic film, *The Ten Commandments*. The massive set, 120 feet high and 720 feet long, was called City of the Pharaoh. After filming was complete, DeMille realized that the set would be too expensive to move, and he didn't want rival filmmakers to use it, so he ordered that it be blown up and secretly buried in a 300-foot-long trench in the dunes. The faux scenery of the Lost City of Cecil B. DeMille lay buried in the shifting sands and forgotten for the next sixty years. A cryptic clue in DeMille's posthumously published autobiography inspired a group of movie buffs to locate remains of the set in 1983. Another thirty years elapsed before archaeologists managed to recover one of the twenty-one fragile, giant 5-ton plaster of Paris sphinxes, which is on display at the Dune Center in Guadalupe.

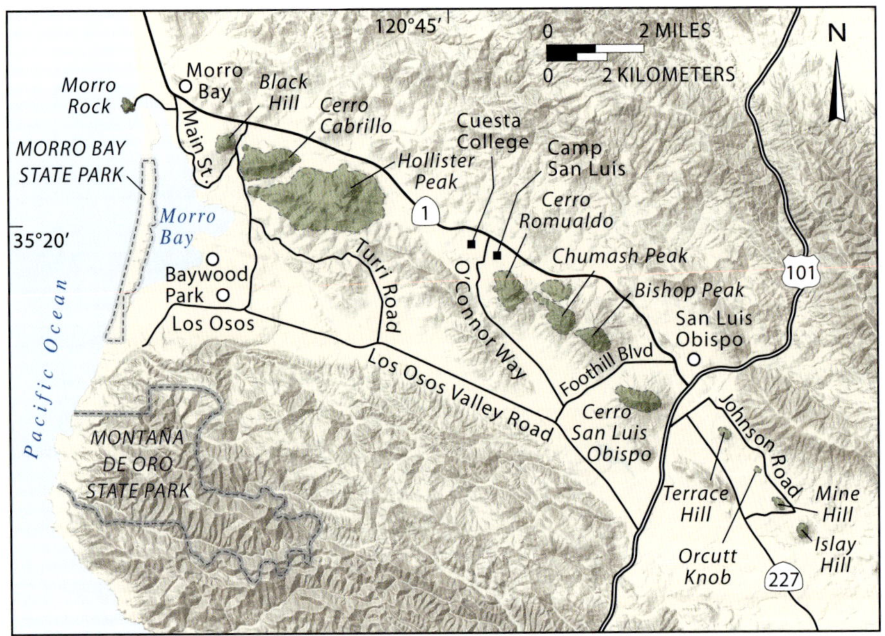

GETTING THERE

Morro Rock and the chain of thirteen volcanic necks, known locally as the Nine Sisters, stretch between Morro Bay and San Luis Obispo. The easiest route to Morro Rock when driving north on CA 1 is via Morro Bay Boulevard (exit 278) into the city of Morro Bay. Follow Morro Bay Boulevard west into town to Main Street, turn right, and then drive north three blocks to Beach Street. Turn left (west) down Beach and go four blocks to Front Street/Embarcadero. Turn right (northwest) onto Embarcadero, drive past the shuttered power plant's three towering stacks, and follow the curve west onto Coleman Drive into the large gravel parking area on the east side of Morro Rock (35°22.14N, 120°51.90W).

VIGNETTE 7

MORRO ROCK AND OTHER KNOBS

Roots of Ancient Volcanoes

SAN LUIS OBISPO COUNTY

Everyone who visits Morro Bay remembers the view of Morro Rock, an isolated knob rising steeply to a rounded top 581 feet above the sea. Describing the rock in 1769, Juan Crespi saw "a great rock in the form of a round morro [promontory or nose] which at high tide is isolated and separated from the coast by little less than a gunshot."

This high dome looks so unusual that we may wonder what it is and how it formed. Was it pushed up from below by some jolly subterranean giant just to befuddle people? Or is it simply a mass of hard and tough rock that resisted weathering and erosion, and whose less durable rock on all sides succumbed? How can we explain its shape, setting, and origin? Let us start with a few hypotheses:

1. Some cataclysmic force rolled it off the mountains to the east to land where it sits today.
2. An earthquake pushed it up eons ago.
3. It is an ancient inactive volcano.
4. It is an erosional remnant of a once larger mass of hard rock, like a sea stack.

To begin our investigation, drive along the shore to Morro Rock for a closer look. Unlike when Juan Crespi was here, high tide no longer turns the rock into an island, and you can safely park your car at its base. While you are in the parking lot, look north to see how Morro Rock is connected to the mainland by a tombolo, an accumulation of sand that built up from the mainland out to the rock. Longshore currents, moving from north to south, created this shoreline feature, and humans have reinforced it with riprap to keep the connection secure.

Then, turn and look south to see the long sand spit that creates and protects Morro Harbor. This stretch of sand was deposited by south-to-north longshore currents. Two jetties protect the harbor entrance and help keep it open.

Aerial view looking southeastward of Morro Rock. Beyond the city of Morro Bay is a chain of thirteen pimple-like volcanic necks that stretches 13 miles east-southeastward to the city of San Luis Obispo and a few miles beyond. —Photograph by John Wiley

Large blocks of rock that have fallen off the east side of Morro Rock have been placed along the west edge of the parking area. Many of these boulders have broken faces that expose fresh rock. Look closely and you'll see that about three-quarters of the rock consists of a fine-grained, gray or blue-gray, homogeneous mass of crystals too small to see with the naked eye. Visible crystals make up the rest of the rock. Scattered throughout are abundant, half-inch-long crystals of a sparkling, light-colored mineral, and a good many other smaller, light and dark mineral grains. The recognizable, well-developed white crystals are the common mineral plagioclase, and smaller, less-abundant black specks are the minerals amphibole and biotite. The kinds, proportions, and shapes of the crystals are typical of the volcanic rock called dacite, a rock intermediate in composition between andesite and rhyolite. The large size of the plagioclase crystals, the fine-grained groundmass, and the circular outline of Morro Rock indicate that this rock probably crystallized fairly quickly at a shallow depth as magma rose upward along a volcanic conduit.

The rock is hard and coherent, which helps it resist erosion. Many surfaces display banding that probably formed when magma flowed and intruded the surrounding host rocks. Several different sets of parallel fractures, called joints, break Morro Rock into regular blocks. The joints are aligned in subparallel sets and at more or less regular intervals. As erosion degrades the rock, blocks tumble to the base of the rock dome.

Morro Rock's equidimensional, nearly circular outline likely reflects the shape of a steep conduit through which magma passed on its way to feed a volcano at the surface. Such a mass is known as an igneous plug or volcanic neck. Eons of erosion have stripped away rock into which the magma intruded and sculpted the volcanic neck into the rounded form we see today. Weathering causes joints to erode more rapidly and deeply than more resistant adjacent rock. Sharp edges and corners have weathered more rapidly than flat faces—hence the phenomena of rounding and smoothing. Nature is the master sculptor.

The fresh surfaces of dacite boulders in the Morro Rock parking area contain white crystals of plagioclase and black crystals of amphibole and biotite. The light-gray blob is a piece of older rock plucked up by the magma when it intruded that older rock. (35°22.12N, 120°51.89W)

Sister Knobs

There is more to the story than revealed just by Morro Rock. It is not convenient to explore northwestward onto the seafloor, so let us head southeast along CA 1, keeping watch for other versions of Morro Rock.

Geologic maps portray at least thirteen separate igneous knobs within a southeast-trending zone that extends nearly 17 miles from Morro Rock to Islay Hill, about 3 miles southeast of San Luis Obispo. This chain of plugs was first mapped in 1904 by the fine early American geologist Harold W. Fairbanks (1860–1952). The intrusions lie along a gently curving, discontinuous arc bowed to the northeast.

A close look at the composition and texture of these knobs indicates that all of them could be siblings of Morro Rock. Although differing slightly in detail, their chemical compositions are so similar that they were probably derived from the same magma source. Their curvilinear arrangement indicates that magma intruded along a single master fracture or a zone of closely spaced subparallel fractures. These fractures, now filled with congealed magma, are called dikes.

Black Hill, a circular, round-topped conical knob 660 feet high, is about 0.5 mile south of CA 1 at its intersection with South Bay Boulevard. A half-mile farther at Quintana Road, an elongate narrow ridge

The northwest side of Hollister Peak, the tallest of the thirteen Nine Sisters. (35°20.4N, 120°42.18W)

south of CA 1, Cerro Cabrillo, rises 910 feet. Still another mile brings you opposite Hollister Peak, with an irregular ragged crest that peaks at 1,250 feet above sea level. It lies at the east end of a large hilly area called Parker Ridge.

Some geologic maps depict Parker Ridge as a single intrusion with an irregular shape; more detailed maps show that it is actually a cluster of many small intrusions that may be the partly exposed top of a single large intrusion. More recent mapping also identifies some small intrusions on the northeast side of CA 1 and a few outcrops of associated lava flows.

Southeast of Parker Ridge, CA 1 passes through nondescript low hills for about 3.5 miles. At the entrance to Cuesta College, the first two of four successive knobs are ahead and south of CA 1: Cerro Romualdo and Chumash Peak. This alignment continues from Bishop Peak to Cerro San Luis Obispo, which lies 0.5 mile west of US 101, about midway between the Marsh Street and CA 1 exits, and then beyond to Islay Hill (elevation 791 feet). People coming into San Luis Obispo from the south on US 101 see Cerro San Luis Obispo ahead and to the north (left), between the Madonna Road and Marsh Street exits.

View looking southeast from Cuesta College to chaparral-covered Cerro Romualdo (elevation 1,191 feet, at right) and Chumash Peak (elevation 791 feet, at left).

Cerro San Luis Obispo looking northwest from Madonna Inn on US 101. (35°16.967N, 120°40.842W)

Why should dikes change upward from thin sheets of igneous rock filling fractures to nearly circular plugs or irregular intrusions? On Iceland and at Kilauea Volcano in Hawai'i, many eruptions begin as linear "curtains of fire" as lava erupts from several long overlapping fissures. Eventually the lava fountains coalesce into a centered eruption as molten rock crystallizing within the fissure concentrates flow into a narrow pipe. Something similar may have happened between Morro Rock and San Luis Obispo.

The fine groundmass of rocks in the line of intrusions suggests that they intruded close to the surface, where they cooled fairly rapidly. Rocks near the surface of the Earth are commonly rich in groundwater. Rising magma that comes into contact with groundwater can produce steam explosions that may blow a hole to the surface, forming a crater and causing broken volcanic rocks to accumulate in and around it. Molten rock may erupt through the hole, forming lava flows. In their book, *Geology Underfoot in Death Valley and Owens Valley*, Robert Sharp and Allen Glazner describe just such events for some of the volcanic craters in the Mono Craters region, particularly at Panum Crater, near Mammoth Lakes. The round neck of Morro Rock may have formed through a similar process.

The line of volcanic rocks was probably generated and intruded as a result of subduction of the Farallon Plate beneath the North American Plate 25 million years ago. Erosion has since destroyed much of the volcanic evidence for this event, leaving us with a string of igneous intrusions, the roots of the volcanoes.

Aerial view looking south of Morro Rock and tombolo connecting it to the mainland.
—Photograph by Cameron Venti on Unsplash

We may conclude that a line of active volcanoes once loomed in the San Luis Obispo area, erupting lava flows, spewing volcanic rubble across the countryside, and filling the sky with volcanic ash. Sedimentary formations in this area contain volcanic debris that could have come from these old volcanoes. Several of the intrusions now stand 1,000 feet above the surrounding terrain of less durable host rocks. The imposing edifice of Morro Rock owes its present form to the erosional forces that stripped away the surrounding, easily erodible rocks and left the tougher volcanic plug behind to stand for all time as a sentinel for mariners and weary travelers.

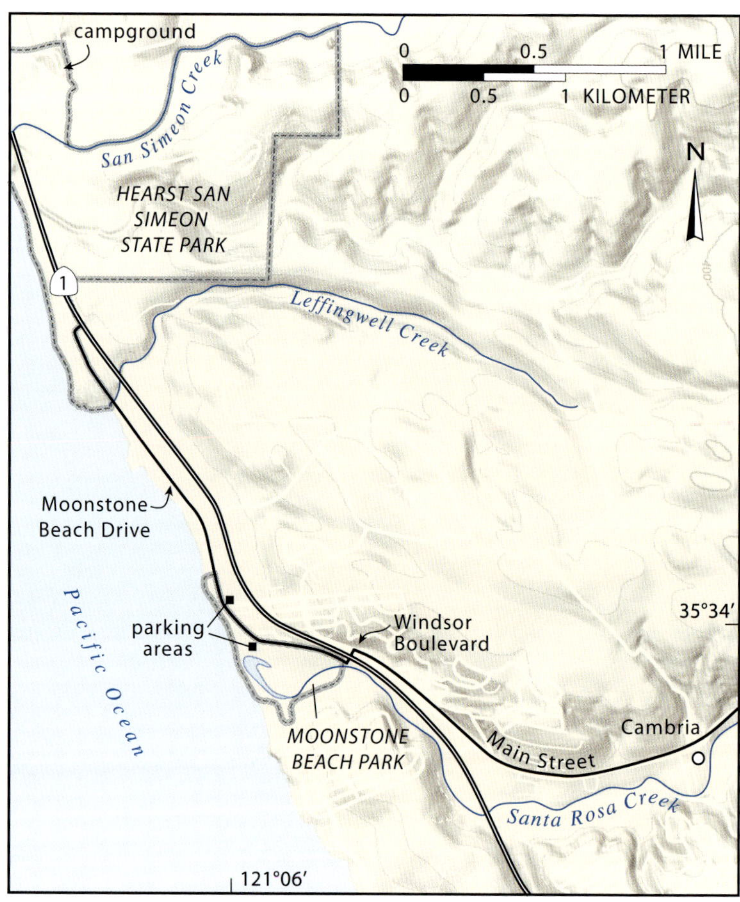

GETTING THERE

For some of the best pebble-hunting sites, turn left (southwest) off CA 1 in Cambria at Windsor Boulevard. Jog seaward 100 feet, turn right (north) and go 0.3 mile on Moonstone Beach Drive to York Street and to a paved state beach parking area on the left (35°34.15N, 121°06.62W). Access to the beach here is easy by way of several walking trails that lead off the mile-long Moonstone Beach Boardwalk. Once on the beach, stroll northwest to where the waterline is close to the sea cliff. Be very aware of occasional rogue waves that prevail here and have swept several beach walkers and pebble hunters to sea, including one of the authors of this book.

VIGNETTE 8

SAN SIMEON BEACH PEBBLES

Sampling the Mysterious Franciscan Formation

SAN LUIS OBISPO COUNTY

Pebble hunting is a bit like fishing—some days are better than others, but a day spent admiring pebbles on the San Simeon coast surely doesn't count against your longevity. Just about everyone—youngsters and oldsters, girls and boys, doctors and lawyers, market speculators, sports fans, and beach bums—loves to look for smooth, lustrous pebbles found on many mixed sand and pebble beaches. All you need is a discerning eye, protection from a cool sea breeze, and a back and legs flexible enough to allow you to pick up stones for a closer look. But beware—inspecting pebbles can be addictive.

The San Simeon coast in central California is famous for its beauty and the variety of its beach pebbles. The best pebble-combing beaches lie between the community of Cambria and San Simeon State Beach Campground at San Simeon Creek. Most high tides, and especially those associated with storms, rework the beaches so conditions change constantly. A good time to look for pebbles is after a major storm. Low tide is obviously better than high tide.

Favored beaches are mostly narrow, with pebbles usually abundant in small coves near the mouths of creeks that bring stones to the shore. Access to most beaches is limited by 25- to 50-foot-high sea cliffs, but several stairways and passable trails lead down the bluff to the beach from clifftop trails or wooden boardwalks.

Beautiful, well-worn, lustrous pebbles 0.25 to 1 inch in diameter can be found on Moonstone Beach (a misnomer explained later), the stretch of San Simeon State Beach extending 0.4 mile north of Windsor Boulevard to Leffingwell Creek. The pebbles lie clustered around driftwood, large rocks, and irregular projections of the sea cliff—obstructions that create centers for unusually high turbulence and eddy currents in wave swash. An abundance of gravel may remain for you to pick through where sand has been winnowed and mostly removed by wave action.

Good pebble hunting can be found on narrow beaches near sea cliffs, such as at Moonstone Beach near Cambria. The black area on the beach is magnetite-rich sand, not oil. (35°34.2N, 121°06.7W)

Pebbles, predominantly reddish-brown chert, on Moonstone Beach near Cambria. Seashell is 2.25 inches long. (35°34.2N, 121°06.7W)

Chert pebbles on Moonstone Beach come in a wide range of colors but mainly reddish brown and green. You can assemble an amazing spectrum of shades of green, ranging from clear apple green through various gray greens to some very dark greens. Other colors include light and dark gray, pale brown, white, and an occasional clear red. In addition to chert, you may find pebbles of dense volcanic and other igneous rocks, and also metamorphic rocks. None of those is as smooth and lustrous as the chert pebbles, and most contain at least a few visible mineral grains.

A little farther northwest on Moonstone Beach is another good place to look for pebbles, mostly reddish-brown chert. Wooden steps lead to the beach a block north of Weymouth Street. Parking is available here (except on crowded weekends) all along the top of the sea cliff, with additional parking about a block farther northwest. The best pebble accumulations on this beach are about 100 to 300 yards southeast of the steps, where shoreline rocks and an irregular sea cliff create turbulence, backwash, and eddy currents during high tides and storm surf.

To find somewhat larger chert pebbles, go to the beach below the parking area at the vista point on CA 1, 0.1 mile south of San Simeon Creek. These stones are mostly 0.5 to 1.5 inches in diameter, rounded, and smooth. Some are bits of rather coarsely crystalline rocks that are not as smooth and lustrous as the chert pebbles on Moonstone Beach. Careful searching, however, will almost always turn up a few nicely smoothed samples of dense chert.

Still larger pebbles, many big enough to qualify as cobbles (larger than 2.5 inches in diameter), lie on the beach at Pico Creek, 1.8 miles north of San Simeon Beach Campground. This spot is accessible from a parking area and a wooden stairway (35°36.85N, 121°08.85W). Look for the beach access sign along the highway south of Pico Creek near the north end of the cluster of motels and restaurants. Scattered among the larger, well-worn, and nicely shaped stones are a few smaller, lustrous pebbles of fine dense chert. North from Pico Creek to San Simeon Village, the beaches are almost entirely sand and are poor places for pebble hunting.

A few of the green San Simeon pebbles are jade, a hard, green, glassy rock composed of the mineral jadeite (a pyroxene mineral, $NaAlSi_2O_6$). Jadeite is one of the most physically durable of all minerals and difficult to break, so it survives the grinding action of a beach environment extremely well. You'll find most jadeite is dull green and hardly gem-like. A few specimens are bright and translucent enough to deserve polishing, but it has been our sad experience that most of the small, green San Simeon pebbles that looked like jade proved, upon being broken, to be only green chert. If they break easily, then you can be sure they are chert.

What you probably won't find at Moonstone Beach are moonstones. Moonstone is the common name for a rare, pearly white, semiprecious stone that belongs to the feldspar group of common rock-forming minerals. It comes in many colors, including white, blue, gray, pink, green, peach, purple, brown, and yellow, but the specimens most eagerly sought are colorless with a pearly sheen. They seem to glow in the sunshine and even more so in the moonlight. This visual effect is caused by the diffraction of light within a microcrystalline structure consisting of stacked, alternating layers of albite feldspar ($NaAlSi_3O_8$) within orthoclase feldspar ($KAlSi_3O_8$).

The Romans thought moonstones were droplets from solidified rays of the Moon. Today some people regard them as a traveler's amulet for protection (especially at night), a passion stimulant, a deflector of negativity, a cure for sleeplessness, and a panacea for a host of other human ailments. Alas, you may search in vain for a true moonstone at Moonstone Beach in the Cambria and San Simeon area; this variety of feldspar is highly uncommon in the bedrock that the pebbles eroded from.

Let's look more closely at chert, the most commonly found pebble on these beaches. Chert is a hard, dense rock composed of silicon and oxygen (silicon dioxide, SiO_2). Silicon and oxygen are, by weight, the two most abundant elements in the Earth's crust and are the building blocks of some of the most common minerals, such as quartz (chert) and feldspar. Specimen quartz, in the form of beautiful six-sided crystals, is commonly found in museums and at gem and mineral shows; specimen chert is usually cut into slabs, polished, and used in jewelry.

Silicon dioxide crystals in chert are so small they are indistinguishable even under microscopic magnification, hence the term cryptocrystalline (*crypto* means "hidden") to describe its texture. The closely intergrown fibrous crystal structure of chert gives it a distinctive homogeneity, a somewhat waxy luster, and a smooth, curved, shell-like fracture habit known as conchoidal fracture.

All minerals made of silicon dioxide (chert, agate, chalcedony, and quartz) are hard and durable—physically and chemically—because exceptionally strong bonds hold its silicon and oxygen atoms together. As a result, substances made of silicon dioxide survive the ravages of weathering and erosion better than most. For example, the beautiful white sand beaches of the western Florida coast around Pensacola consist almost entirely of quartz grains that have survived hundreds of millions of years of weathering and physical battering before being delivered to the beaches by rivers. Nearly all other minerals have deteriorated, leaving only the quartz. Being as tough as quartz, San Simeon chert pebbles are also survivors.

Chert's toughness and homogeneity have made it a useful and valuable commodity throughout human history. Ore processing mills at many early mining operations were ball mills that used chunks of chert, tumbling in a cylinder with the ore, as grinding balls to pulverize the ore and release metals like gold, silver, and copper. In today's world, compounds of silicon are the building blocks upon which transistors and microchips are built—the heart of our modern digital devices.

Native Americans fashioned arrowheads and spear points from chert. They prized the chert for its hardness, durability, attractive colors, and, most of all, its fracture habit, which permitted them to cleave sharp edges. Native Americans in the West found most of their chert as nodules or thin irregular layers enclosed within limestone beds that are widely exposed in the western United States (see Vignette 18). We know that calcium carbonate ($CaCO_3$), the building block of limestone, accumulates on the seafloor, so the chert must have originated there, too.

Why are San Simeon chert pebbles so delightfully varied in color, texture, and luster? It is because they come from an unusual and complex bedrock source. They are not the average run-of-the-mill beach stones, and their exceptional beauty and character reflect the unusual origin of their parent rocks. Most of these pebbles come from the Franciscan Formation, a rock assemblage that is present in large areas in the hills inland from the San Simeon coast.

Some cherts in the Franciscan Formation are colorful, mainly thin, regular layers called bedded or ribbon cherts—rocks that were deposited on the deep-ocean floor. Seawater contains considerable dissolved silica and calcium carbonate. Clear ocean water teems with minute microorganisms of various kinds, including forams, which build their tiny shells out of calcium carbonate, and simple single-celled plants (diatoms) and marine protozoa (radiolaria), which use the dissolved silica to build siliceous skeletons. When these animals die, their remains sink, ever so slowly, to the seafloor where they accumulate just as slowly in huge abundance as opaline ($SiO_2 \cdot nH_2O$) siliceous ooze. Calcium carbonate builds up and forms limestone in shallow water, but calcium carbonate dissolves in the deep ocean, so all that is left at those depths is the siliceous ooze. It goes through several stages of consolidation before becoming ribbon chert.

The Franciscan Formation was one of the major mysteries of California geology for many years. Like most rock units, its name derives from a geographic locality, in this case, the San Francisco Bay area. It is more widely exposed in northern and central California than in the southern part of the state. The Franciscan Formation is exposed not only in the sea cliffs north of San Simeon Creek, but also offshore of southern California (see Vignette 3).

The abundance of chert pebbles tells us that the Franciscan Formation contains a lot of chert, but how did these once deep-sea cherts come to be part of the Franciscan Formation? Before initiation of the San Andreas fault at this latitude—about 16 million years ago—the North American Plate overrode an oceanic plate known as the Farallon Plate. Islands, deep-sea sediments, and parts of the underlying oceanic crust were scraped off the top of the down-plunging Farallon Plate by the edge of the North American Plate, in the way a snowplow scrapes and piles up snow in front of it. This ocean bottom debris accumulated at the western edge of the North American Plate to form the beginnings of the California Coast Ranges. Much of this accreted and mildly metamorphosed sedimentary debris is now known as the Franciscan Formation. Younger sedimentary and volcanic rocks were later deposited over the Franciscan Formation and then uplifted to create the Coast Ranges of today.

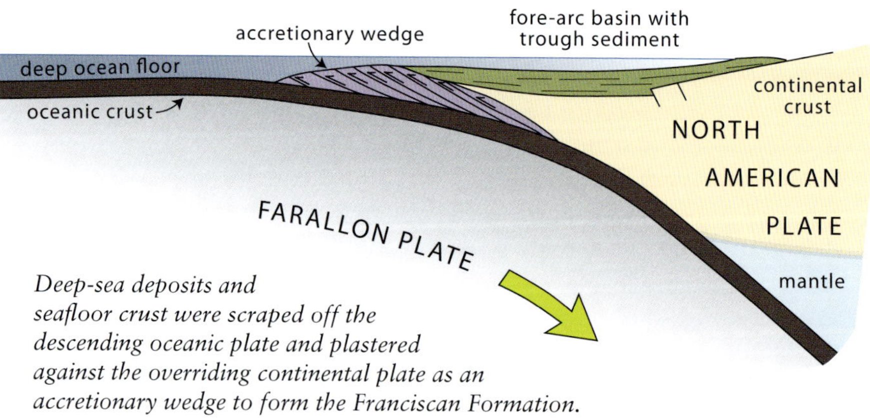

Deep-sea deposits and seafloor crust were scraped off the descending oceanic plate and plastered against the overriding continental plate as an accretionary wedge to form the Franciscan Formation.

After Franciscan cherts became part of North America, they were uplifted within mountains that rose near the coast. Weathering and erosion ultimately exposed them on the surface and eventually broke them into pieces that slope-wash, streams, and rivers carried down to the seashore. There, wave action tumbled the stones about, leaving behind polished pebbles that now cover the beaches. If these pebbles were interrogated, what a yarn they would spin.

VIGNETTE 9

THE WHITTIER NARROWS

A Boon to Communications

LOS ANGELES COUNTY

Most drivers never recall driving up and over any hills when they pass through the Puente Hills uplift on their hurried and harried way south on Interstate 605 from the San Gabriel Valley for a day at the beach or a long wait in line at Disneyland. The Puente Hills are a formidable east-west geographic barrier; the Whittier Narrows is the principal passageway through this barrier.

In the recent geologic past, the San Gabriel River cut down about 800 feet through the Puente Hills, forming the Whittier Narrows. Erosion and landsliding widened the narrows to its present width of about 2 miles. The narrows divide the Puente Hills uplift into two parts: the Repetto Hills to the west and the Whittier and Chino Hills to the east. I-605 passes through the Whittier Narrows, along with other roadways, railroads, pipelines, power lines, flood-control channels, and lines of communication. How did the San Gabriel River breach the Puente Hills to form this vital urban lifeline?

The answer is not unusual by classical geologic standards. The river was in place before the Puente Hills rose. Known as an antecedent river, it pre-dates the modern landscape. This scenario is likely to happen only in regions where the Earth's crust is actively deforming. Some of the deformation (folding and faulting) in the Puente Hills is so young that it is likely the hills are still rising. In fact, two earthquakes originating deep beneath the Whittier Narrows in early October 1987 indicate that the region is tectonically active today. Epicenters of those earthquakes, magnitudes 6.1 on October 1 and 5.4 on October 4, plot along the general trend of the Whittier-Elsinore fault zone. This zone parallels the Puente Hills and continues into the populous San Gabriel Valley and northern Los Angeles Basin.

Before the Puente Hills rose, however, the San Gabriel River and several smaller streams flowed south out of the San Gabriel Mountains and established shallow channels across their floodplains. Then a low ridge, the first inkling of the future Puente Hills, started to

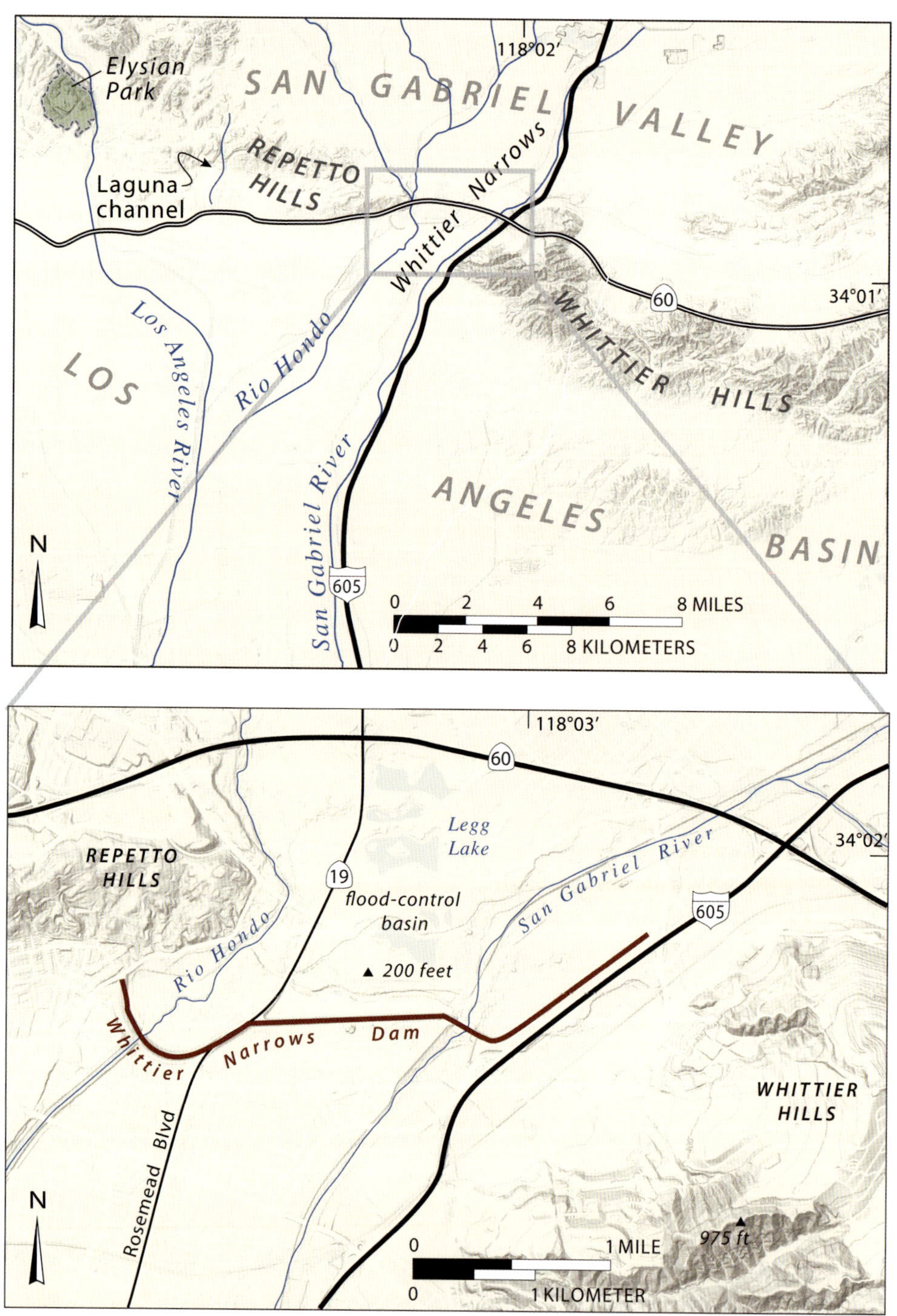

GETTING THERE

The Repetto and Whittier Hills are part of the Puente Hills uplift, which separates the Los Angeles Basin from the San Gabriel Valley. Interstate 605 passes through the uplift by way of the Whittier Narrows water gap.

OIL IN THE LOS ANGELES BASIN

A thick sequence of marine sedimentary rocks was deposited in the Los Angeles Basin from about middle Miocene to latest Pliocene time (about 15 to 3 million years ago). Intense shortening deformation of these rocks over several million years has created isolated groups of hills, separated by broad valleys, like the Puente Hills. This deformation has been a major factor in the concentration of significant petroleum accumulations within folds and against faults, such as in the folded Montebello oil field west of the Whittier Narrows, and to the southeast along a fault that passes through Whittier and La Habra.

rise across their courses. Some of the smallest streams were probably turned aside immediately, but larger, more powerful streams, including the San Gabriel River, were able to erode their beds as rapidly as the ridge rose, cutting ever-deeper channels across the growing barrier.

Most of these streams were unable to challenge the rising ridge and so were forced to find new courses. Some of these deflected streams turned east to become tributaries of the San Gabriel River, increasing its flow and allowing it to erode its channel even more easily. In the end, only the San Gabriel River survived, continuing to cut down into and across the Puente Hills, as an antecedent stream. And the saga goes on: as the hills continue to rise, the river continues to erode its way down as it flows south to the sea. If a rising ridge exceeds a river's ability to continue to cut downward, then the river must divert and carve another channel. The water gap then converts to a wind gap, the name given to an abandoned river course.

Water gaps and wind gaps are not unique to southern California. They abound in other parts of our country, particularly in the Appalachian Mountains, where the Cumberland Gap is a classic example. There, as here, such gaps allow relatively easy passage through otherwise formidable ridges and mountains. Many water gaps, especially in the eastern United States, were created as rivers cut down through masses of hard rock, which subsequently became imposing granitic and metamorphic ridgelines. Here, the San Gabriel River cut through softer rocks of the Puente Hills.

On clear days, people looking south from tall buildings in Pasadena can see wind gaps in the crest of the Repetto Hills. Four of these gaps are important transportation routes. From west to east, these

Aerial view in 2018 of the Whittier Narrows. The confined bed of the San Gabriel River extends obliquely from the upper right to the lower center. —Google Earth image

are the Laguna Channel (Long Beach Freeway [I-710]); Coyote Pass (Monterey Pass Road); a relatively wide unnamed channel (Atlantic Boulevard); and a little farther east, another unnamed pass (Garfield Avenue). Although the smaller streams that cut those passes finally lost their battles against the rising Puente Hills, we can thank them for their partly successful fight. These wind gaps make it easier to pass over the Puente Hills and to understand the origin of these saddles in the Puente Hills.

Rio Hondo is a curious anomaly within the Whittier Narrows. It is a seasonal (wet-weather) stream with at least two branches that drain parts of Pasadena-Alhambra and Monrovia-Arcadia. Rio Hondo follows a course through the Whittier Narrows a mile or so west of the San Gabriel River. Upon emerging from the south end of the narrows, Rio Hondo turns southwest to join the Los Angeles River between Downey and South Gate. Its artificially controlled flood-control channel still follows that route. Why Rio Hondo preferred this roundabout course to the ocean by way of the Los Angeles River is a mystery. Perhaps the large amount of sand and gravel carried by the San Gabriel River raised its channel sides (overbank deposits) just enough that the Rio Hondo would have had to flow uphill to join it.

Rivers do most of their work (erosion, transportation, and deposition) during flood stage. Huge river-laid deposits of sand and gravel, rich in cobbles and boulders, were deposited by the San Gabriel River south and west of Azusa. The size and weight of these rocks attest to the enormous energy that was needed to move such large objects and cut the Whittier Narrows at the same time. The river's floods must have been frightful deluges before dams and huge flood-control impoundments robbed the river of its strength. The Whittier Narrows Dam and flood-control basin, completed in 1957, were designed to capture flood water from both the San Gabriel River and Rio Hondo. The flood-control basin is normally empty parkland.

The next time you whisk through the Whittier Narrows on Interstate 605, tip your hat to the San Gabriel River and its hard-won victory in its long battle with the steadily rising Puente Hills. The San Gabriel River deserves a bronze plaque somewhere in the Whittier Narrows to celebrate the convenience it gives to the people of the greater Los Angeles area.

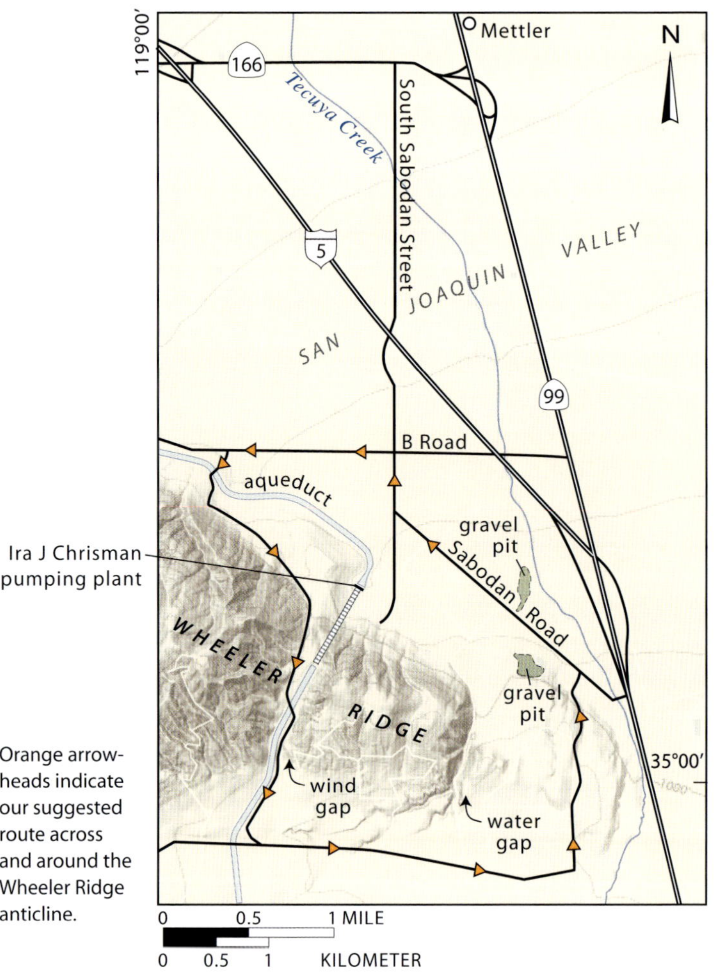

Orange arrowheads indicate our suggested route across and around the Wheeler Ridge anticline.

GETTING THERE

To get a close look at Wheeler Ridge, exit Interstate 5 at CA 166 (exit 225). Go east about 1.2 miles on CA 166 (Maricopa Highway) to South Sabodan Street, then turn right (south). Seven-tenths of a mile after crossing over I-5, turn west (right) on a wide, paved, but unsigned street (B Road). Continue for 1 mile, turn left onto the first paved road to the south, and then cross the bridge over the California Aqueduct. Continue about 1 mile southeast between the aqueduct and the power line and follow the big curve uphill and south across a cattle guard to the aqueduct crossing at the crest of the wind gap. Park in the wide area (35°00.353N, 118°59.227W) on the near (west) side of the aqueduct.

If traveling north on CA 99, exit at CA 166 (exit 225), go west to Sabodan Street, turn south and follow the instructions in the paragraph above.

VIGNETTE 10

WHEELER RIDGE

Gaps across a Rising Fold

KERN COUNTY

Wheeler Ridge is clearly visible to motorists on two of southern California's busy freeways, Interstate 5 and CA 99. The junction of these two major highways north of Grapevine Canyon lies 0.5 mile north of the east end of Wheeler Ridge, so thousands of people pass the ridge every day. It is a fine place to visit in springtime when the grass is green and wildflowers grace the smooth slopes south of Wheeler Ridge.

Views are best for southbound travelers on I-5 and reasonably good on southbound CA 99. A little careful sidewise look to the southwest by northbound motorists on either route reveals the essential features. In either direction, motorists should have no trouble spotting the four large pipes rising 518 feet from the Ira J. Chrisman Wind Gap Pumping Plant on the north side and near the east end of Wheeler Ridge. (The plant is named after the chairman of the California Water Commission from 1967 to 1976.) The pipes cross the ridge through a wide-open saddle 500 feet below the ridge crest. The saddle is a wind gap, the abandoned valley of stream that formerly flowed across the ridge. The wind gap is an invaluable gift to the California State Water Project, also known as the Edmund G. Brown California Aqueduct.

West of and separated from Wheeler Ridge (elevation about 2,100 feet) are the subdued, smooth grassy slopes of the Pleito Hills (elevation about 4,600 feet). On the skyline beyond, the rugged San Emigdio Mountains top out at Tecuya Mountain (elevation 7,163 feet). Wheeler Ridge is a separate entity, rising more than 1,000 feet above the smooth apron of sediments in the San Joaquin Valley that slopes gently northward toward Bakersfield (elevation about 400 feet). So, what is the ridge doing out there all by itself, a sort of isolated maverick? At first blush, it sort of looks like a mega landslide.

Wheeler Ridge may be isolated, but it is no maverick or landslide. It is one of the best examples anywhere of an upward-arching fold, called an anticline. It is so young that the shape of the ridge replicates

View looking south from South Sabodan Road of aqueduct pipes and road using the saddle in Wheeler Ridge left by an abandoned stream. The floor of the San Joaquin Valley fills the middle ground, with the San Emigdio Mountains on the distant skyline.

the arching form of the fold—modified only slightly by rill wash and erosional gullies on its side slopes. Two water gaps and several wind gaps form saddles in the anticline's crest. The formation of water and wind gaps here is similar to that in the Ventura Avenue anticline (Vignette 4) and the Whittier Narrows (Vignette 9), where rivers cut across growing folds. If a stream still flows through the gap, it is called a water gap.

Wheeler Ridge stretches 8.5 miles from east to west with a maximum width of 3.5 miles. Like the curved back of a breaching whale, it dives beneath the ground at its east end and blends into the Pleito Hills to the west. Oil and gas companies have drilled many holes in Wheeler Ridge anticline in search of oil and gas, which commonly rise from source rocks below and may become trapped in the apex of an anticline (see figure on page 46). These holes and others in the nearby Tejon and North Tejon oil fields reveal that the subsurface geologic structure is considerably more complex than a simple surface anticline would suggest. In fact, Wheeler Ridge anticline is the upper plate of the active, south-dipping Wheeler Ridge thrust fault, a major structure that lies along the entire base of the north side of Wheeler Ridge. A thrust fault is a gently inclined fracture surface along which the upper block (hanging wall) has been pushed upslope relative to the lower block (footwall). At greater depths, other thrust faults in older rocks further complicate the anticline's interior.

A Round-Trip over and around Wheeler Ridge

The directions at the beginning of this vignette will get you to a parking area in the wind gap on the west side of the California Aqueduct. There, you will see boulders littering the nearby hillslopes. They weathered out of the Pliocene to Pleistocene Tulare Formation, a sedimentary deposit of unconsolidated coarse sand, gravel, and boulders laid down by streams as recently as 250,000 years ago. The Tulare Formation is the youngest unit in the sequence of folded sedimentary strata in the Wheeler Ridge anticline. This is a significant bit of information because it lets us infer that the entire Wheeler Ridge has been folded and uplifted in the last 250,000 years. Geologically speaking, that's yesterday! Wheeler Ridge is a most active geologic structure.

Although bedding layers in the Tulare Formation are crude and a little hard to see, you can probably tell that they dip gently down to the south. The anticlinal crest lies several hundred feet north of the parking area, so backtrack by car or foot about 0.3 mile down the road to a large gully on the west side of the road. Look at the bare slopes and cliffs at the gully's head to see the change in direction of bedding inclination that defines the crest of the anticline.

View looking west from a point on the aqueduct wind gap road (35°00.636N, 118°59.144W) at the folded beds in the Tulare Formation. Dashed white lines trace the anticlinal form of the fold.

Turn your car around safely by continuing north approximately 150 yards to a graded flat in the curve at the bottom of the hill. Retrace your route back up the hill to the first cattle guard and the west side of the aqueduct. Cross the aqueduct at the crest of the wind gap and continue south.

Boulders in the Tulare Formation are worth a closer look. A representative collection lies at the base of the hillslope across from the aqueduct, about 50 feet beyond the second cattle guard and on the east side of the road. They are hard igneous and metamorphic rocks, with an occasional chunk of marble, derived from distinctive rocks in the San Emigdio Mountains on the southern skyline. These rocks may also be studied in an embankment of boulders along the north side of B Road.

This boulder-rich part of the Tulare Formation consists of coarse debris that floods swept out of the San Emigdio Mountains and Pleito Hills before Wheeler Ridge existed. Most of the boulders are rather angular, their sharp edges blunted ever so slightly, indicating that they could not have traveled very far. Maximum boulder size in this vicinity is about 6 feet. It can be difficult to imagine the immense amount of energy required to transport nearly 9-ton boulders to their current resting places. These deposits differ hardly at all from sand, gravel, and boulders in the beds of streams that drain out of the San Emigdio Mountains today.

From the crest of the aqueduct wind gap, you may make a roundtrip down the south flank of the anticline by circling to the east, up and over the ridge again and back to South Sabodan Road. Continue 0.3 mile south on the potholed road to a Y intersection; take the right fork and continue for another 0.3 mile to a T intersection. Turn left and drive eastward away from the aqueduct. Proceed 2 miles on the once-paved but now unmaintained road and beyond a large oil field operations yard to an unnamed graded gravel road. Turn left onto this road, which is a bit rough but not hazardous. It will take you over the anticline between the unnamed water gap, 1 mile to the west, and the narrow, active Tecuya Creek water gap on the right (east), along the west edge of highway US 99.

The 2-mile-long unnamed road over the anticline emerges onto South Sabodan Road about 300 yards east of a large aggregate quarry from which Tulare Formation sand and gravel are extracted. The hard, fresh boulders from the Tulare Formation make excellent commercial gravel when crushed.

Northbound travelers may turn left onto South Sabodan Street and head toward the entrance to the Ira J. Chrisman pumping plant and then north to either freeway. Travelers wishing to go south on I-5 can turn southeast (right) on South Sabodan Road at the quarry, drive southeast 0.3 mile to the frontage road paralleling the freeway, and continue south about 1.5 miles to the Tejon Commerce Center at Laval Road, and to an on-ramp for I-5.

Water and Wind Gaps

Up to now, we have been too close to Wheeler Ridge to see much of the features along its crest. As can be seen from the air and on topographic maps, several other shallow wind gaps are present in the Wheeler Ridge crest, west of the California Aqueduct. Initially, several streams flowed northward out of the San Emigdio Mountains and Pleito Hills across a gently sloping alluvial apron to the floor of the San Joaquin Valley. Then, the Wheeler Ridge anticline started to arch across the paths of these streams. Smaller streams and those crossing the most rapidly rising part of the ridge were diverted eastward away from their channels and around the nose of the growing fold after they cut only shallow notches across it. These abandoned stream valleys became wind gaps as the anticline continued to rise.

The stream that cut the aqueduct gap began carving into the uplifting ridge about 60,000 years ago. It crossed the younger and less rapidly rising east end of the anticlinal fold, where it could cut down as rapidly as the anticlinal barrier rose across its path. Water from some of the diverted streams may have augmented its flow, focusing erosion there. The "aqueduct stream" maintained its course across the anticline during the first half of Wheeler Ridge's rise until it could no longer keep pace. Perhaps the ridge began to rise faster than the stream could erode its bed, or possibly the increasing width of the fold defeated the stream. Whatever the reason, the rising fold blocked the stream and forced its water eastward through a large, unnamed water gap, which is being abandoned in favor of Tecuya Creek even farther to the east. The prominent wind gap now occupied by the California Aqueduct is the largest and most spectacular wind gap on the crest of Wheeler Ridge.

In the mid-twentieth century, aqueduct engineers seeking an inexpensive means of crossing Wheeler Ridge at a convenient elevation spotted the wind gap. It was at the right elevation for water to flow east by gravity to the huge Edmonston pumping plant where aqueduct water is boosted 1,926 feet to cross the Tehachapi Mountains at the ground surface. Use of the Wheeler Ridge wind gap capitalized on the stream's erosional work as it battled the rising anticline. Aqueduct

Aerial view looking northwest across the unnamed water gap in the Wheeler Ridge anticline. Dashed lines highlight the anticlinal structure. The creek flows north from the lower left to the upper right of the image. Automobile (circled, lower left) and road provide the scale.

engineers should periodically honor that valiant stream, gone now but not forgotten.

Young (late Pleistocene and Holocene), tilted stream deposits at the east end of Wheeler Ridge suggest that the fold is rising at the rate of at least 3 millimeters per year, comparable to the uplift rate of the Ventura Avenue anticline (Vignette 4). At that rate, it would have taken only about 200,000 years to raise the entire Wheeler Ridge from start to finish. On the other hand, the hypocenter (origin at depth) for the 1952 M7.5 Kern County earthquake was directly under Wheeler Ridge. Surveys made half a year after the 1952 earthquake found that Wheeler Ridge had risen as much as *2 feet* since the previous survey made just one year earlier. That uplift suggests that the fold grows in height and length by herks and jerks, not simply a few fractions of an inch year after year. Thankfully, the 1952 uplift happened prior to aqueduct construction. Nevertheless, engineers must keep a wary eye on the amount, location, and current rate of uplift. Even small changes in elevation could play havoc with aqueduct gradients.

VIGNETTE 11

CAJON PASS

San Andreas Fault Paves the Way for I-15

SAN BERNARDINO COUNTY

Thousands of vehicles and people pour through Cajon Pass on Interstate 15 every day. Three rail lines, four power lines, communication lines, and several gas and petroleum pipelines also use the pass. Cajon Summit, the high point on the highway at 4,260 feet, is about 2 miles northwest of the true pass that the railroads utilize. Motorists have a good chance of seeing a many-engine train climbing or descending the pass. In earlier days, Native Americans, explorers, Spaniards, and Mormon pioneers used this path between the desert and the coastal regions.

Were it not for Cajon Pass, I-15 and the railroads would need to climb a steep 11 percent grade to at least 5,000 feet above sea level, a scant 5 miles from the south base of the mountains near San Bernardino. But the highway crosses the summit of the pass 13 miles from the south base, with a grade averaging 3.25 percent, and the longer railroad grade is only 2 percent. A unique combination of circumstances created this gateway through the formidable Transverse Ranges, so Cajon Pass is a gift from nature that merits our attention and appreciation. Why is it here?

After climbing a long grade, I-15 approaches the summit of the pass by climbing steep, southwest-facing slopes that Levi F. Noble (1882–1965), a prominent early geologist who first mapped this area for the US Geological Survey, called the Inface Bluffs. Beyond the summit, the highway starts its long, gradual, much gentler northward descent toward Victorville, 17 miles ahead. On the skyline to the south and southwest, the rugged crests of the San Bernardino and San Gabriel Mountains, attractively snowcapped in winter and spring, rise over 10,000 feet. North of the high mountains and south of the Inface Bluffs is a broad, open amphitheater of much lower elevation and gentler relief called Cajon Valley. Sloping gently northeastward from the Inface Bluffs is a smooth surface of coalesced alluvial fans of early and middle Pleistocene age (1.9 million to 250,000 years before present), hereinafter called the Victorville apron. Its south margin and broad washes terminate abruptly and anomalously where they are beheaded

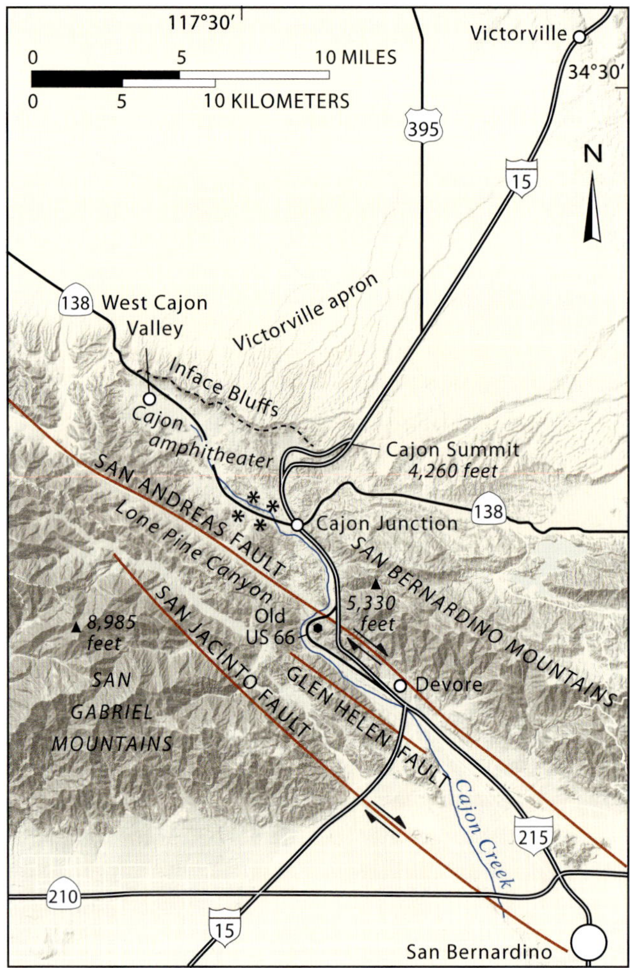

Stars indicate the location of the Mormon Rocks.
The solid circle locates Blue Cut.

GETTING THERE

Cajon Pass is located along I-15 between San Bernardino and Victorville. You can appreciate this vignette as you zoom over the pass, but you can also get off the interstate to check out a few interesting geologic features. To reach Blue Cut, exit I-15 northbound at Kenwood Avenue (exit 124), proceed a few hundred yards west to the bottom of the hill and turn right onto Cajon Boulevard and follow it north. To reach Mormon Rocks Interpretive Trail, take exit 131 for CA 138 and head to the Mormon Rocks fire station 1.5 miles northwest from the interstate.

Aerial view looking west along the crest of the Inface Bluffs. The Victorville apron slopes north (right) from Forest Route 3N24 on the crest of the bluffs to Victorville in the Mojave Desert. The heads of arroyos flowing down this slope from their sources in the San Gabriel Mountains are beheaded at the bluffs' crest.

at the crest of the Inface Bluffs. Two anomalous characteristics of the apron help us piece together the puzzling aspects of Cajon Pass.

Gravel in the Victorville apron contains numerous pebbles and cobbles of a dark gray, slabby mica schist called the Pelona Schist, which is widely exposed in the San Gabriel Mountains to the southwest. The schist is a thinly foliated metamorphic rock rich in mica. These and other distinctive rock fragments in the apron must have come from those mountains, but how did they cross the low ground of Cajon Valley? There is no existing downhill route the cobbles could follow from the San Gabriel Mountains to the Victorville apron. That is anomaly number one.

An aerial overview reveals that the upper reaches of the Victorville apron are incised by broad washes—arroyos—some 100 feet into the apron surface. Streams commonly build smooth alluvial fans with a slope nicely adjusted to carry their water and sediment load; they do not ordinarily erode a channel into their own fan without some independent perturbation such as a change in climate or a rise or increased tilt of the land.

Alluvial deposits of the Victorville apron contain Pelona Schist (dark-gray cobbles) and other rock types derived from the San Gabriel Mountains. The orange field book is 7.5 by 4 inches. (34°22.852N, 117°32.673W)

Extremely precise measurements of ground surface elevation changes from repeated highway surveying over many years show that this region has risen at a rate of about 0.5 foot per century during the period of the surveys relative to the country to the north. It is reasonable to suppose that the land has been rising for a longer time, and that its rise allowed the washes to dissect the Victorville apron.

The washes are peculiar in other ways, too. If you were to walk up one of them, such as Manzanita Wash about 4 miles west of I-15, you would find that it extends with full width and depth right to the brink of the Inface Bluffs. There it disappears; its headwaters are gone. It is a beheaded wash as are the other washes. That is anomaly number two.

The best clues to explain these anomalies and the origin of Cajon Pass involve the history of Cajon Creek, which lies between the San Gabriel and San Bernardino Mountains. These ranges are part of the Transverse Ranges, so called because of their westward trend, transverse to the general northwest trend of the mountains elsewhere in southern California. These west-trending mountains form a nearly continuous wall breached by Cajon Pass, which connects the San Bernardino Valley on the coastal lowland and the Mojave Desert.

Cajon Creek and I-15 enter the southern margin of the mountains at Devore, where I-215 merges with I-15. Most streams flowing out of

Aerial view looking north at the steep, south face of the Inface Bluffs (heavily shadowed) and the incised surface of the Victorville apron. The dashed line traces the course of Manzanita Wash, now beheaded at the crest of the Inface Bluffs and sloping across the apron to Victorville in the hazy distance.

View looking south from the crest of the Inface Bluffs across Cajon Valley to the crest of the San Gabriel Mountains on the skyline. The San Andreas fault lies along the north edge of the mountains.

the mountains follow courses roughly perpendicular to the mountain front, but Cajon Creek and its next-door neighbor to the west, Lytle Creek, flow southeastward, more parallel than perpendicular to the mountain front. Both creeks follow the trends of swarms of fractures that parallel the San Andreas fault. In its lower part, Cajon Creek lies midway between the San Gabriel and San Bernardino Mountains, and between the San Andreas and San Jacinto faults; it follows the Glen Helen fault. Streams are masters at finding easy rock to erode, and they rarely miss crushed and fractured rocks in a major fault zone.

As Cajon Creek canyon penetrates the mountains, it becomes narrower and deeper, and I-15 rises higher onto their northeast wall. The San Andreas fault lies a scant mile or two northeast of I-15 along the base of the San Bernardino mountain front, with a bearing that gradually converges with the highway. About 3 miles northwest of Devore, where Cajon Creek has been displaced right-laterally by the San Andreas fault, Old US Highway 66 and the creek curve broadly to the north and northeast, respectively, and enter stream and highway roadcuts carved in dark-gray Pelona Schist, a rock formation that is a major part of the terrain southwest of the San Andreas fault. The schist resists erosion, so Old US Highway 66 and the creek pass through it in a steep-walled canyon called the Blue Cut (34°16.0015N, 117°27.534W), named for the bluish-gray color of the schist.

The change in the bearing of Old Highway 66 puts it on a collision course with the San Andreas fault. The highway and fault meet at the east end of the Blue Cut. A quick look to the west, under favorable lighting and visibility, gives a fine view up the long, narrow, abnormally straight Lone Pine Canyon (Vignette 19), aligned and eroded directly along the trace of the San Andreas fault zone. Lone Pine Canyon is anomalous in that it cuts across drainages flowing northeastward out of the high San Gabriel Mountains to the southwest. From the air the canyon looks like a discordant scar hacked into the face of the Earth by a giant machete. Levi Noble was so impressed with the physiography of the canyon that he called it a rift like the great rift in Africa.

At Cajon Junction, where I-15 and Cajon Creek meet CA 138, the interstate enters wide open Cajon Valley. Your attention will be drawn almost immediately to prominent ridges of pale tan sandstone outcrops, mostly north of CA 138 and west of I-15. These ridges are locally known as the Mormon Rocks to memorialize early Mormon pioneers who camped along Cajon Creek on their way from San Bernardino, which they founded, to Utah.

Bold, north-dipping outcrops of the Cajon Formation sandstone on the northeast side of Cajon Creek in Cajon Valley are called the Mormon Rocks. (34°19.11N, 117°29.55W)

Take the opportunity to study the geologic and cultural history of Cajon Pass by walking Mormon Rocks Interpretive Trail, an easy 1-mile-long loop. It begins at the Mormon Rocks fire station located on CA 138, 1.5 miles west of I-15. The rocks are sandstone and pebbly conglomerate assigned to the Cajon Formation of early to middle Miocene age. Look closely and see that the pebbles are volcanic and slaty rock types, not Pelona Schist. Note how the rock layers dip northward, and some are cross-bedded with cross-bed orientations that indicate sediment transport to the southwest. Geologists would conclude that the sediment was derived from an area of volcanic and metamorphic rocks to the northeast, in the Mojave Desert, transported southwest by a river system, deposited in a basin and lithified about 18 to 13.5 million years ago, and then tilted northward. These sedimentary rocks lie on the northeast side of the San Andreas fault.

The San Andreas fault and Cajon Creek are major players in how Cajon Valley became an easy passage through the mountains. In geologic parlance, the San Andreas is a right-lateral strike-slip fault; the blocks on opposite sides slip past each other horizontally, and no matter which side you are on, the opposite side moves to the right. The fault separates the North American and Pacific Plates, no trivial function. These two plates make up about one-third of the surface area of the Earth, and crossing Cajon Pass means crossing from one of them to the other.

Geologic evidence indicates that total horizontal displacement on the San Andreas fault in southern California is at least 100 miles, with the rocks west of the fault moving northwestward. The total displacement may amount to as much as 160 miles. Displacement along the San Andreas fault caused the 1857 Fort Tejon earthquake and the more disastrous 1906 San Francisco earthquake, both with magnitudes approaching 8. It also caused the 1812 earthquake whose rupture passed through Cajon Pass and terminated at the headwaters of Lone Pine Canyon.

The San Andreas fault extended into this area about 6 million years ago, before any of the present topography existed. The cross section and map for Stage A show the topographic setting as it existed roughly half a million years ago. By that time, the south face of the mountains rose steeply from the floor of San Bernardino Valley to a rugged crest that was probably more, and perhaps considerably more, than 5,000 feet above sea level. The divide between drainages to the coast and desert lay along this crest. Streams from high on the desert

View north across Mormon Rocks and the broad Cajon Valley that Cajon Creek eroded between the San Gabriel Mountains and the Mojave Desert. The Inface Bluffs along the skyline reveal interiors of beheaded alluvial fans of the Victorville apron.

face of the mountains spread a veneer of sand and gravel, the Victorville apron, over the 18-to-13.5-million-year-old Cajon Formation beds in the area now occupied by Cajon Valley. Because this alluvium was derived from the mountains, it contained fragments of their rocks, particularly chunks of Pelona Schist.

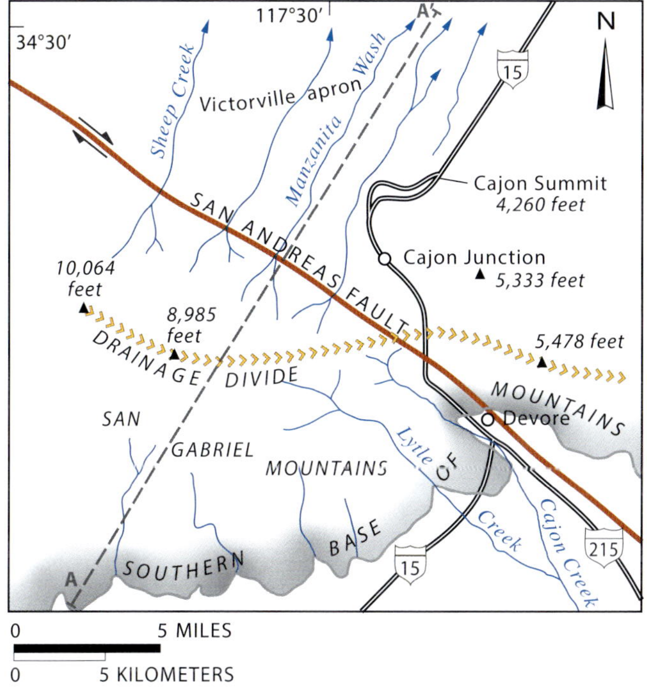

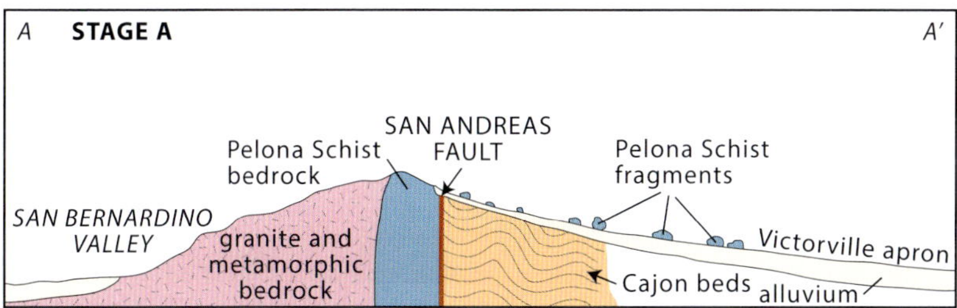

Stage A. Easily erodible Cajon Formation strata have slipped into position along the San Andreas fault (see cross section), but they are buried by a veneer of alluvial gravel containing Pelona Schist stones and other rocks from the San Gabriel Mountains. The Victorville apron extends northward from the foot of the mountains. The drainage divide, located along the high peaks of the mountains, separates streams that flow ultimately southward to the Pacific Ocean from those that flow into the interior of the Mojave Desert.

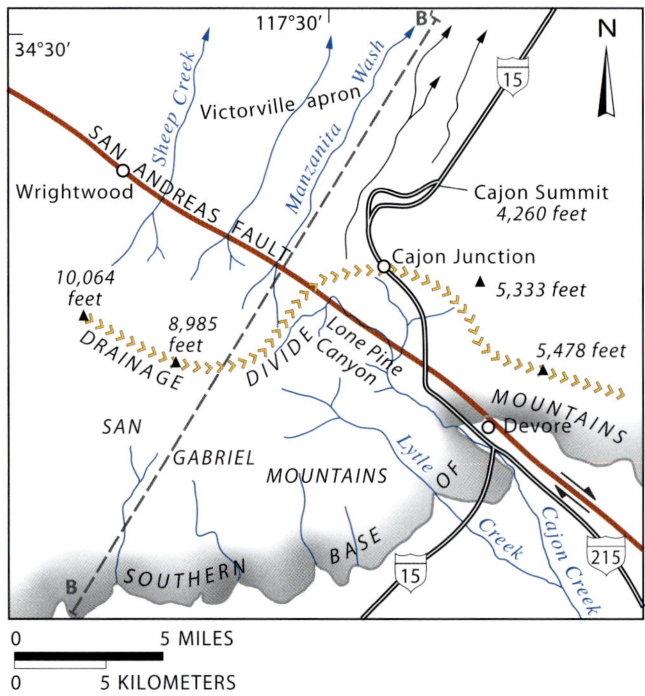

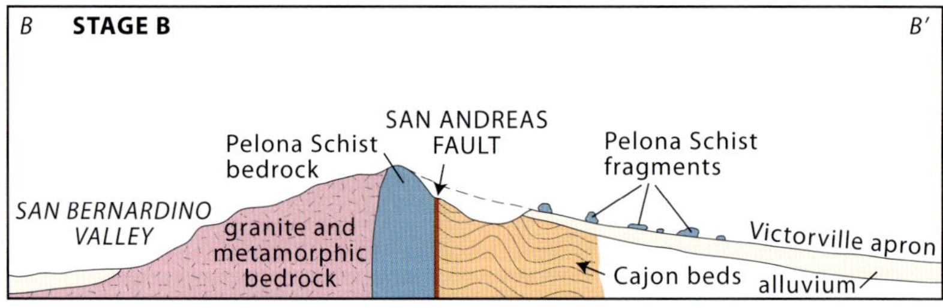

Stage B. Cajon Creek crosses the San Andreas fault and erodes Cajon Formation beds, thus shifting the drainage divide northward. Lone Pine Creek, a tributary to Cajon Creek, works headward along the San Andreas fault zone. At least one Victorville apron stream is beheaded.

Anyone attempting to cross the mountains from the coastal side before Cajon Pass formed would have had to climb about 4,000 feet up the steep south face of the mountains to an elevation above 5,000 feet before dropping to the head of the Victorville apron for the smooth passage northeast to Victorville. Someone traveling in the opposite direction would have had to climb at least 1,000 feet higher than the floor of the Mojave Desert, and then would have faced a very steep descent to the San Bernardino Valley. It would have been like crossing the rugged

San Bernardino Mountains today near the community of Crestline (elevation 4,600 feet), about 12 miles southeast of Cajon Pass.

While streams flowing from the northeast side of the mountains were building the Victorville apron, Cajon Creek was slowly eating into the mountains from the south. By the time Cajon Creek extended 5 miles into the mountains, the San Andreas fault lay only about 1 mile farther, so like any prudent explorer, Cajon Creek sent out advance scouts—tributaries. One tributary extending northeast discovered the

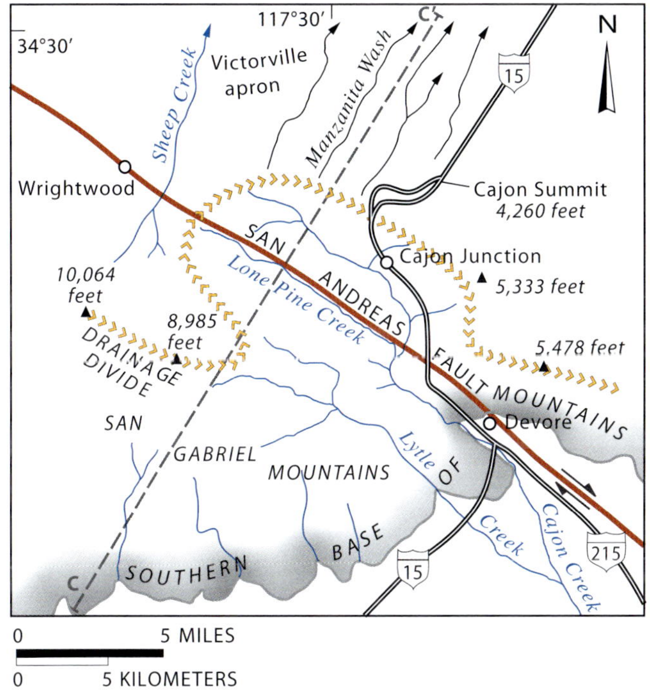

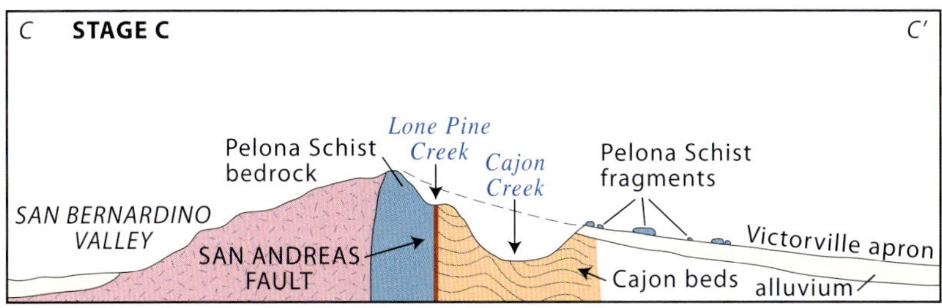

Stage C. Cajon Creek enlarges the erosional valley it is creating in the Cajon Formation beds. With the help of its parallel tributary, Lone Pine Creek, the drainage divide shifts farther to the northwest. Incipient Inface Bluffs are probably forming. Manzanita Wash and other arroyos are beheaded.

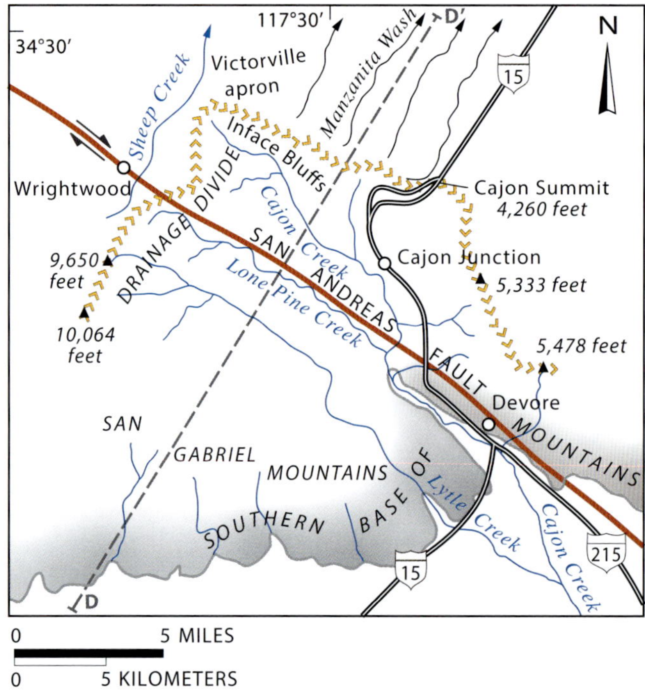

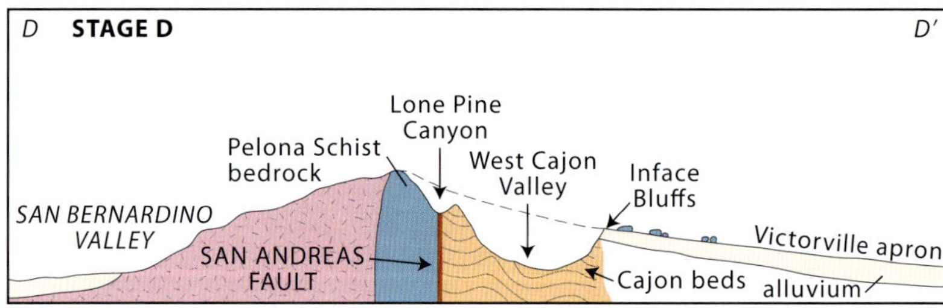

Stage D. Continued erosion by Cajon Creek and its tributaries carves Cajon Valley and the Inface Bluffs as they exist today. The drainage divide has retreated 5 miles northward, and the crest of Cajon Pass has been lowered from over 5,000 feet to 4,260 feet. Only Sheep Creek continues to flow directly from the San Gabriel Mountains across the San Andreas fault onto the Victorville apron, still carrying Pelona Schist cobbles and pebbles. Today, Cajon Creek and its tributaries drain to the Pacific Ocean, whereas all creeks and streams north of the drainage divide empty into the interior of the Mojave Desert.

wide zone of crushed rock within the San Andreas fault zone and so cut across it into the easily erodible Cajon Formation sandstone. This tributary rapidly eroded its valley northwestward along the fault and soon began to capture the headwaters of streams flowing northeast from the mountains into the desert. These captures so augmented

Cajon Creek's water supply that the tributary quickly became the main trunk of Cajon Creek.

Once Cajon Creek found the Cajon Formation sandstone, its tributaries voraciously ate into it and eroded headward, creating the Inface Bluffs. Gullies eroded the bluffs' steep south face, and streams and groundwater sapped their base, thus gnawing into the Victorville apron and causing the bluffs to retreat northeastward. This erosion drove the drainage divide to the northeast and also lowered it because the apron slopes downward in that direction. The arroyos on the apron were cut off from their former sources of water and rock debris in the San Gabriel Mountains. The southwest ends of the arroyos were left hanging in empty air at the Inface Bluffs, literally beheaded. Recession and lowering of the drainage divide at the brink of the Inface Bluffs continue slowly today. The cross sections, maps, and their captions in the accompanying illustrations should help you to visualize this sequence of events.

An explanation for the two anomalies mentioned earlier now becomes clear. The arroyos on the Victorville apron once extended several miles farther southwest into the northeast slope of the San Gabriel Mountains. They have been beheaded by the erosional attack of Cajon Creek tributaries, thus producing anomaly number two. As for anomaly number one, the Pelona Schist fragments in the Victorville apron were carried from the mountains and spread across the fan and into the Mojave Desert before Cajon Valley existed. These anomalies, while relatively subtle items, provide important clues to the succession of events that created Cajon Valley and Cajon Pass.

Although erosion by Cajon Creek is the agent most directly responsible for carving the valley, it could never have done the job if easily erodible sedimentary rocks had not already been moved into place by displacement along the San Andreas fault. Thus, the San Andreas fault created the critical relationships that Cajon Creek exploited.

So, the next time you escape crowded southern California by way of Cajon Pass for a vacation in the Sierra Nevada, the desert, or a weekend in Las Vegas, give the San Andreas fault a friendly nod in passing. It is your transportation benefactor as well as your seismic nemesis.

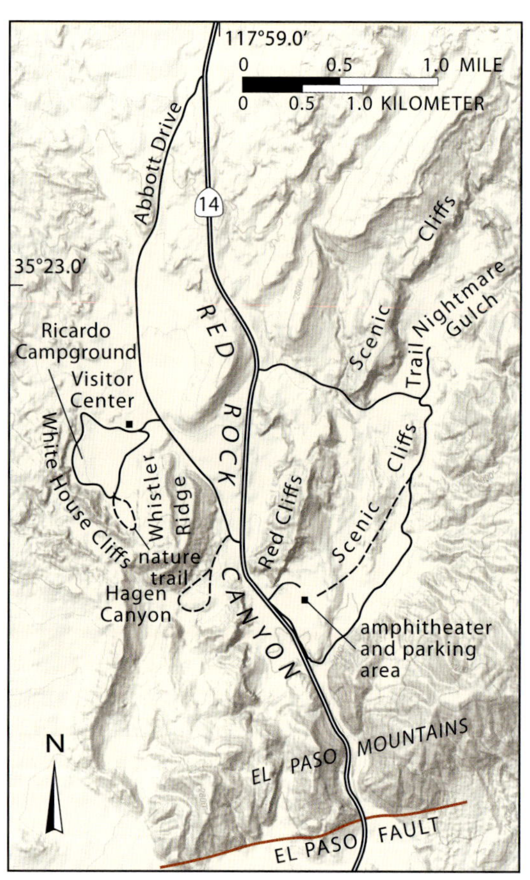

GETTING THERE

CA 14 passes through the heart of Red Rock Canyon 25 miles northeast of Mojave.

VIGNETTE 12

RED ROCK CANYON

Finding Faults in a Geologist's Training Ground

KERN COUNTY

Every day, thousands of motorists see the brightly colored badland cliffs of Red Rock Canyon in the Mojave Desert as they speed north on CA 14 from heavily urbanized southern California to vacation areas of the eastern Sierra Nevada. You may have seen the cliffs in numerous movies and TV shows including *Jurassic Park* and *Bonanza*. The cliffs also provided a castellated background for cowboy films, including *Red Rock Canyon* with Ronald Reagan.

Geologists enjoy the scenery, too, but they treasure the canyon and its rocks even more as an archive of information about the history and evolution of the surrounding countryside. To them, the canyon is a library, its sedimentary beds being like pages in books. Reading those pages requires knowledge, skill, and experience. Let's start at the beginning and read the story.

CA 14 goes northeastward from the town of Mojave across some 20 miles of the broad and gently sloping alluvial apron that flanks the south end of the Sierra Nevada. The highway curves broadly north and heads directly toward the abrupt south face of El Paso Mountains: the scarp of El Paso fault. The fault is nicely exposed at the mountain base a few hundred yards east of the highway, where pink, 7-to-5 million-year-old, readily erodible sedimentary rocks to the south are faulted against orange-hued, resistant Mesozoic bedrock to the north.

The road soon enters a narrow gorge cut into dark-brown, extremely fractured, reddish-stained Mesozoic granite and older metamorphic rocks intruded by the granite. After passing through nearly 1 mile of granite, the road bursts dramatically into the Red Rock amphitheater, with its striking badland cliffs. Badlands are bad only in the sense that their steep and barren slopes make them useless for grazing and difficult to cross. All of us tend to detour around them, but the scenery of badlands is invariably striking, commonly colorful, and irresistible, so we stop and admire them. Bryce Canyon and

View looking east across CA 14 to an exposure of El Paso fault at the mouth of Red Rock Canyon. (35°20.794N, 117°57.9W)

Staircase badlands topography in layers of the Dove Spring Formation in Hagen Canyon of Red Rock Canyon State Park. (35°21.9N, 117°59.8W)

Cedar Breaks in Utah and the Badlands of South Dakota are good examples. Smaller badlands dot the arid Southwest, and many are protected as county or state preserves. Red Rock Canyon's badlands are part of Red Rock Canyon State Park, established in 1970.

Badlands form most commonly in sedimentary deposits that are unconsolidated enough to erode easily but coherent enough to stand in very steep cliff faces. A key component of badlands is how erosion-resistant beds alternate with thicker, less resistant ones, thus forming the staircase-like appearance. The resistant beds protect the less resistant beds beneath, which erode into pillars that endure, for a while, until the protective roof rock is eroded back.

For leisurely inspection, turn right from CA 14 into a large parking area with restrooms, 1.25 miles from the gorge entrance. Along the north edge of the parking area are the Red Cliffs, noted for their closely spaced, vertical drainages that make them look almost like drapery pleats, especially toward the west end.

At first glance the sedimentary layering looks horizontal, but it actually dips northward, away from you, at an angle of about 15 to 20 degrees. Closer inspection of the cliffs reveals that the various layers are mostly sandstone, siltstone, claystone, and gravel mixed with volcanic ash beds. Some of the ridges are capped with dark-gray, erosion-resistant basaltic lava flows. What you see here is a small part of the 1-mile-thick Dove Spring Formation. Many of the bright-white beds are volcanic ashes that fell when these sediments were being deposited.

Rain splash and surface runoff are the dominant processes of erosion for carving a geometric collage of columns, spires, flutes, chutes, knife-edge ridges, niches, alcoves, and sundry other forms. They are highly effective and powerful processes, especially where heavy rains fall on barren slopes that shed surface runoff in rills and gullies. Rill and gully channels can be so deep, steep-sided, and narrow that walking into one is like walking into the fold of a gigantic set of curtains; it can be dark in there on a sunny day, as if you were in a cave.

Variations in color, including white, cream, beige, brown, pink, red, and green, reflect variations in the amount and oxidation state of iron in the different layers. Iron that has been oxidized at the surface is missing three electrons and is bright rusty red, whereas the original iron in the rocks is mostly missing two electrons and imparts a dark, typically greenish hue.

Near the base of the cliffs, the ground is littered with small pebbles weathered out of the sedimentary rocks. Most are fragments of various volcanic rocks, but a few are granite or metamorphic rocks.

These stones contribute to the spectrum of colors in the cliffs, but the large content of volcanic ash contributes more. Look for some small bright-white chips of rock on the ground. Many of them have black flecks of mica in them. Try placing a rock chip on your tongue. If it's ash, it may stick to your tongue because the former volcanic glass in the ash has weathered to clay that can absorb lots of water.

Walk south across the parking area, away from the Red Cliffs, until you can see their top and the thick, massive layer of pink rock that caps them. It is highly fractured and weathers to subdued knobby surfaces. CA 14 passes through this layer about 900 feet north of its intersection with the road into the parking area. There you can lay hands on the bluff-capping layer. Close up, you see that a fine matrix of volcanic ash encloses small angular fragments of mostly volcanic rocks. This is an ash-flow tuff, which was deposited as a fast-moving mixture of hot gas, incandescent ash, and pumice erupted from a nearby volcano 12 to 13 million years ago. Such ash flows are deadly where they roar down inhabited mountain slopes, such as happened to Pompeii in the AD 79 eruption of Mount Vesuvius. The volcano that produced the ash-flow tuffs in Red Rock Canyon is now 40 miles

Massive, pink volcanic ash-flow tuff caps Red Cliffs on the north side of the amphitheater parking area, 300 yards east of CA 14. (35°21.85N, 117°58.67W)

Close-up view of ash-flow tuff, a volcanic rock consisting of pumice (white), ground-up pumice (pink matrix), and fragments of green, gray, red, and tan volcanic rocks. Pencil for scale.

east of Red Rock Canyon and south of the Garlock fault, having been shifted there by left-lateral displacement on the Garlock fault.

As you leave the parking lot and turn north onto CA 14, look at the pink outcrops on the right where the road starts to curve right. A cluster of small faults, inclined about 45 degrees east, displaces the contact between the ash-flow tuff and the underlying sedimentary beds by about 30 feet.

To see more of Red Rock Canyon, turn west (left) onto Abbott Drive, 0.3 mile north of the amphitheater parking area. Follow Abbott Drive about 1 mile to the park ranger station, visitor center, and campground. The visitor center offers exhibits, as well as maps, publications, and information on many aspects of the park's human and natural history.

A short nature trail starts from the campground road near the last group of campsites at the base of White House Cliffs, a few yards beyond campsite 46 (35°22.08N, 117°59.62W). Follow the broad trail south, up the ridge to the saddle and wooden bench at cairn 12 (35°21.944N, 117°59.498W), and look out into the Mojave Desert (north) and Fremont Valley (south). One of the several branch trails leads south into Hagen Canyon, another goes west to promontories that overlook Ricardo Campground, and a third one goes east up a steep ascent to the top of Whistler Ridge for a splendid panoramic view of Red Rock Canyon and larger parts of the Mojave Desert.

Faults, center, inclined to the right, displace the contact (dashed line) between the massive pink ash-flow tuff and the underlying sedimentary layers. View is 200 feet east of CA 14 and 900 feet north of the Red Cliffs. (35°21.79N, 117°58.84W).

Look north from the top of Whistler Ridge to see northwest-tilted layers of the Dove Spring Formation, consisting of rusty-red, cross-bedded conglomerate and sandstone; white mudstone and volcanic ash; massive pink ash-flow tuff; and thin, gray layers of mudstone. Along the crest of Whistler Ridge is a resistant basaltic lava flow that overlies and protects the less resistant sedimentary layers beneath from erosion. Fossils in these strata and isotopic dating of the basalt indicate that the rocks are Miocene in age, deposited between 19 million and 7 million years ago in an inland basin. The sandstone and conglomerate strata are stream deposits, and the mudstone was deposited in river floodplains and lakes. Nearby volcanoes intermittently erupted ash, cinders, and lava flows that ended up in the basin.

Several other trails are worth exploring in Red Rock Canyon State Park. An easy and well-marked trail follows a 1-mile loop in Hagen Canyon, starting from a spacious parking area at the south end of Abbott Drive. From it, you can wander into several side canyons and see other exposures of the massive pink bed of ash-flow tuff.

View looking north at west-tilted basaltic lava flow ridges that cap sandstone and siltstone in the Dove Spring Formation as seen from the west flank of Whistler Ridge. Red Rock Canyon visitor center is in the left center of the image.

A slightly more demanding walk northeast out of the Red Cliffs parking area follows a broad trail, actually an abandoned road, that parallels the Red Cliffs. In a little less than 1 mile it meets an active road and enters a broad valley that ends in a splendid amphitheater 1 mile farther north at the Scenic Cliffs area. A vehicle road into the Scenic Cliffs area branches from CA 14 a bit more than 1 mile north of the road into the Red Cliffs parking area. No sign identifies this junction, but it is easy to spot where a straight reach of CA 14 bends into a broad curve to the west. The sinuous and unmaintained condition of the road into the Scenic Cliffs area, however, makes it easier and more fun to walk the trail from the Red Cliffs parking area. The round-trip hike of 4 miles is well worth a half day, even a full day. This is one of the best parts of Red Rock Canyon, scenically and geologically, and remote enough to lend a sense of adventure. Bring plenty of water!

A network of old roadways, foot trails, and flat-floored washes makes wandering and exploration easy. Besides its engaging badland scenery, this area displays faults cutting beds in the cliffs; massive

The rubbly slope is a landslide of pink ash-flow tuff boulders along the Scenic Cliffs trail. (35°22.21N, 117°58.41W)

landslides, especially involving the pink ash-flow tuff in both the Red and Scenic Cliffs; and some dark basalt dikes that cut through the Cudahy Camp Formation. The traveled vehicle road ends near the mouth of Nightmare Gulch, a narrow defile with steep walls and a favorite among off-road vehicle enthusiasts for testing their driving skills.

Geologic History

Red Rock Canyon has fascinated geologists ever since one of the greatest of the American breed, Grove Karl Gilbert (1843–1918), gave it a complimentary nod in an 1875 report on the geology of the western United States. Discovery of fossil mammal bones increased interest during the early 1900s. Scientific papers on the area continue to appear regularly.

Most large regions can be conveniently subdivided into geologic provinces, based on their unique landforms and geology. Southern California has nine such provinces. Three of them—the Mojave Desert block, the Sierra Nevada, and the Basin and Range—come to a triple junction at Red Rock Canyon. This area had a ringside seat at happenings within those three provinces, especially during the last 19 million years.

The Mojave Desert block is shaped like a wedge, with a sharp western point at Lebec that broadens eastward. The San Andreas fault bounds its southwest side; the Garlock fault zone marks the north side. El Paso fault is part of the Garlock fault zone. The Sierra Nevada block trends north and south, west of the Sierra Nevada frontal fault system,

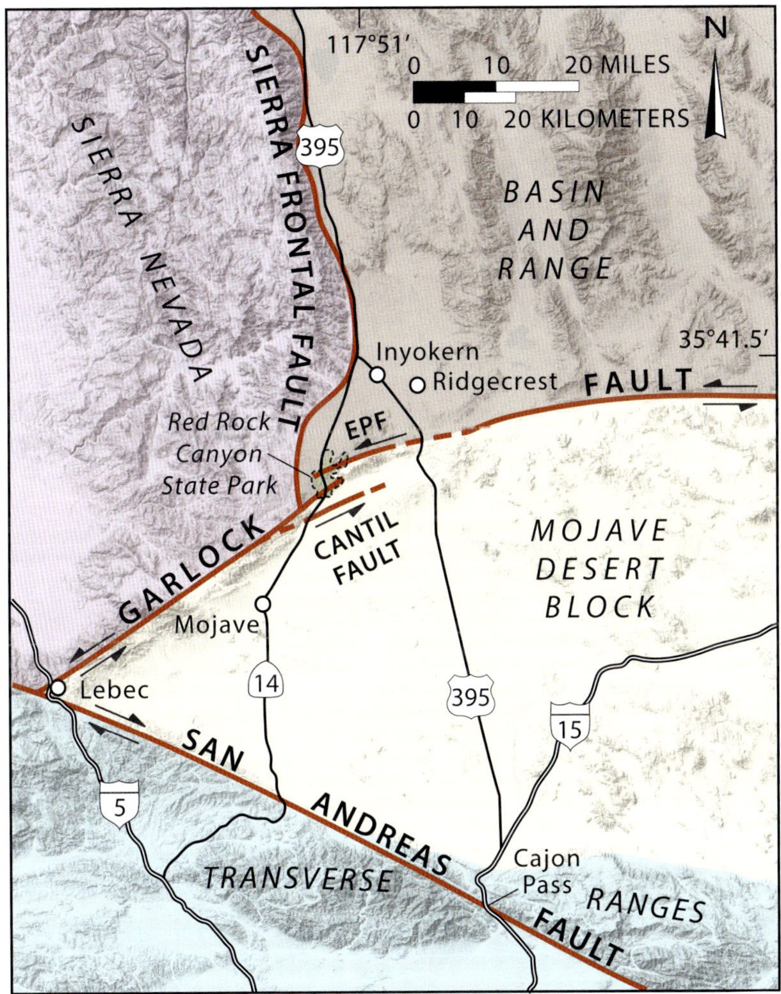

Tectonic setting of Red Rock Canyon at its strategic location at the triple junction of the Sierra Nevada, Basin and Range, and Mojave Desert geologic provinces. The Garlock fault slices through a down-dropped crustal block—a graben—bounded by El Paso (EPF) and Cantil faults. I-15 crosses the Transverse Ranges and San Andreas fault in Cajon Pass.

which passes along the west side of the Red Rock Canyon area and terminates against the Garlock fault. Red Rock Canyon lies within the southern part of the Basin and Range province, which contains a succession of lofty mountain ranges that trend north, separating deep basins such as Owens, Searles, Panamint, and Death Valleys. In the 1880s, geologist Clarence E. Dutton described the parallel mountain ranges of the Basin and Range as "looking upon the map like an army of caterpillars crawling northward out of Mexico." Rocks exposed in Red Rock Canyon record events that occurred within these three provinces, especially along the Garlock and Sierran fault systems.

Tectonism is still quite active in the region, as evinced in 2019 by the Ridgecrest earthquakes of July 4 (M6.4) on a left-lateral strike-slip fault subparallel to the Garlock fault, and July 5 (M7.1) on a right-lateral strike-slip fault. Both of these earthquakes occurred on the Naval Air Weapons Station China Lake near the town of Ridgecrest, about 15 miles northeast of Red Rock Canyon.

During early to middle Tertiary time, 55 to 19 million years ago, the Mojave Desert block was eroding and shedding great quantities of sediment south and west into adjacent marine sedimentary basins. Near the end of this period, however, the geologic gremlins in charge of the Mojave Desert block apparently decided it was unwise to export all that debris into the sea; some should be kept at home. Perhaps they recognized that erosion is the great eraser, and deposition the great preserver, of geologic records. Small basins, where sedimentary records of events could be preserved, formed onshore by downwarping and faulting within the Mojave Desert block. The Barstow Basin (Vignette 13) is one of these; El Paso Basin at Red Rock Canyon is another.

GARLOCK FAULT

Garlock fault, named for a former mining community that is now a ghost town about 10 miles east of Red Rock Canyon, is one of the longest faults in southern California, stretching nearly 160 miles from Frazier Park near Lebec and I-5 to the southern end of Death Valley. Its horizontal displacement is about 40 miles, *left-lateral*, in contrast to most other faults in California, including the San Andreas, which slip *right-laterally*. The Garlock fault cuts off the south ends of mountains and valleys in the Basin and Range and separates them from the Mojave Desert with imposing vertical displacements of several thousand feet.

AGE	ROCK UNIT	THICKNESS	DESCRIPTION
QUATERNARY	older alluvium (1 million years to present)		alluvial and lake silt, clay, and gravel
LATE MIOCENE	Dove Spring Formation (12.5 to 8 million years)	5,400 feet	stream and lake conglomerate, sandstone, siltstone, mudrock, rhyolite tuff, with interbedded basaltic lava flows
EARLY MIOCENE	Cudahy Camp Formation (19 to 14 million years)	1,000 feet	andesite, tuff, coarse clastic rocks
EOCENE / PALEOCENE	Goler Formation (60 to 50 million years)	7,900 to 13,050 feet	nonmarine conglomerate, sandstone, mudstone
PALEOZOIC	Garlock assemblage		metamorphic rocks

Generalized stratigraphy of rock formations in the western El Paso Mountains. Wiggly lines indicate unconformities.

Sedimentary layers that accumulated in El Paso Basin from 17.1 million years to something less than 8 million years ago make up Dove Spring and Cudahy Camp Formations. Their cumulative thickness of 6,400 feet does not mean that a basin of that depth formed first and then filled to the brim. Rather, the basin sank slowly, bit by bit, and sedimentation kept pace, so the basin was always nearly full. El Paso Basin accumulated an even greater thickness of sediment, 60 to 50 million years ago, the 13,000-foot-thick Goler Formation. Those layers were tilted and then eroded before the younger formations were deposited. The angular unconformity thus created marks a 35-million-year hiatus between deposition of the Goler and Cudahy Camp Formations.

Most of the Red Rock Canyon area is made up of Dove Spring Formation rocks, which are simply the top of a very thick sedimentary sequence that filled El Paso Basin. It comprises 5,400 feet of stream and lake deposits, including conglomerate, sandstone, mudstone, and chert, as well as four distinct basaltic lava flows and numerous layers of pink tuff.

Among the layers of upper Dove Spring sediment are fossil reminders of a time when fertile grasslands and open forests surrounded the shoreline of a shallow freshwater lake. Remains of a rich fauna of mammals include more than a half dozen species of early three-toed horses, four species of camels, mastodons, two rhinoceroses, wild dogs, pronghorn antelopes, deer, saber-toothed tigers, cats, weasels, squirrels, chipmunks, hedgehogs, moles, shrews, mice, rabbits, wolverines, foxes, and possibly a skunk or two. The grazing must have been good. Although grasslands probably predominated, Dove Spring strata also contain fossils of pinyon pine, locust, cypress, acacia, and palm trees. The annual rainfall may have been around 15 inches, three times the canyon's present quota but similar to that of present-day Los Angeles in a good year. Even so, can you imagine rhinos and saber-toothed tigers wandering around southern California today?

In 1988, two young geologists, Dana Loomis and Douglas Burbank, published an exhaustive study of the Cudahy Camp and Dove Spring Formations. Using as reference points the ages of lava flows and volcanic ash beds within the sedimentary sequence, they created a detailed timescale for the two formations. They estimated the rate of sinking of El Paso Basin and the rate of deposition within it, both of which varied with time. They also measured past reversals in the Earth's magnetic field as recorded by magnetic minerals within sedimentary layers of the Dove Spring Formation. They obtained ages of the rock layers by matching the local record of magnetic reversals against a master chart compiled from worldwide analyses of seafloor sediments, where the full magnetic record is preserved.

Loomis and Burbank concluded that Cudahy Camp sediments began to accumulate about 19 million years ago and continued for 5 million years, when deposition was interrupted for about 1.5 million years. During this hiatus, the Cudahy Camp beds were tilted and eroded, creating the surface upon which the Dove Spring Formation was deposited. The Dove Spring Formation accumulated 5,400 feet of strata from 12.5 to 8 million years ago.

Loomis and Burbank used just about every known geologic technique to determine the sources of the sediment dumped into the slowly subsiding El Paso Basin. Cudahy Camp sediments came primarily from the Mojave Desert block directly to the south. The formation also contains sediment eroded from the underlying Goler Formation,

---

NO COLLECTING

All natural, geological, and archaeological features in Red Rock Canyon State Park are protected by law, including fossils. No collecting is allowed except by scientific permit approved in advance by the California Department of Parks and Recreation. Look instead for fossils displayed at the park's visitor center.

---

as well as from 300-million-year-old Permian sedimentary rocks and the 470-million-year-old Ordovician Mesquite Schist. Much of the debris eroded from the Mojave Desert block is volcanic ash and fragments from dark lava flows and other volcanic rocks. The sources of the Cudahy Camp rocks are no longer nearby because that terrain has shifted left-laterally along the Garlock fault. Instead, the source area is now about 40 miles east of Red Rock Canyon in the Eagle Crags area on Naval Air Weapons Station China Lake. Earlier vertical displacement certainly occurred along the Garlock fault in earliest Miocene time, but the separation of Dove Spring beds across the fault tells us that horizontal displacement started only 10 million years ago.

Dove Spring sediments also came initially from the Mojave Desert. Although largely volcanic, the sediments came from a greater variety of sources than did the Cudahy Camp Formation. More pieces of relatively old granite and metamorphic rocks suggest that erosion had stripped the volcanic cover from some areas, exposing the older rocks beneath. More than twenty ash beds from sources unknown lie among sedimentary strata throughout the Dove Spring Formation, especially near its base, as seen in the colorful cliffs alongside CA 14.

About 8 million years ago, the proportion of granitic debris in the Dove Spring sediments increased dramatically, and its character changed from that of granite in the Mojave Desert to that of the Sierra Nevada. Evidently, at that time, uplift along the southern part of the Sierra boundary fault began to elevate the Sierra Nevada, an event geologists had wanted to date for a long time; Loomis and Burbank finally did it in 1988. The Sierra still shed sediments into and across the Red Rock Canyon area until a few thousand years ago. In fact, as you drive north from Red Rock Canyon State Park on CA 14, look to the southwest (over your left shoulder) as you pass the "3,000 elevation" or "Dove Springs Off Highway Vehicle Area" signs; you can see a gently sloping surface stretching from the foot of the Sierra to the edge of Red Rock Canyon. This is a remnant of a sediment transport surface for eroded debris from the Sierra across the Red Rock region into depressions that straddle the Garlock fault in the Mojave Desert a few miles south of Red Rock Canyon.

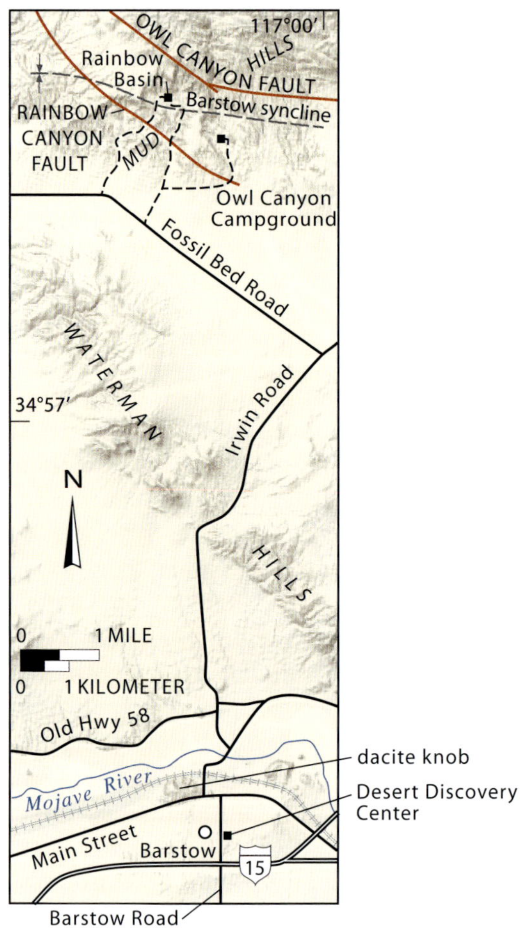

GETTING THERE

People coming from the Los Angeles area should take Barstow Road (CA 247, exit 183) north from I-15. Three-tenths of a mile north of the freeway is the BLM Desert Discovery Center (831 Barstow Road), a Bureau of Land Management facility that houses exhibits about the high-desert environment and provides information about road conditions. The center also displays the Old Woman meteorite, a 2.75-ton chunk of nickel and iron that was found in 1976 in the Old Woman Mountains about 100 miles east of Barstow. It is the largest meteorite ever found in California, the second largest in the United States, and well worth seeing.

One mile north of the freeway, Barstow Road ends at a T intersection with Main Street. Turn west (left) and go 0.2 mile to First Avenue. Turn north (right) on First Avenue, following the signs to CA 58. Cross the Mojave River and railroad yard via an impressive steel truss bridge, then curve east (right) past the Harvey House train station and the extensive

VIGNETTE 13

RAINBOW BASIN

Geologic Laboratory along the Calico Fault

SAN BERNARDINO COUNTY

We can think of three good reasons to pay a visit to Rainbow Basin and Owl Canyon in the Mud Hills. First, they are wonderfully colorful—favorites of artists and photographers—where the layered sedimentary rocks crop out in shades of bright green, white, red, and brown. Second, they are a key link in our knowledge of mammals that lived during middle Miocene time. Vertebrate paleontologists named an interval of geologic time the Barstovian Stage (16 to 12 million years ago) after the fossils found here. Third, the rocks in Rainbow Basin are folded and faulted in a manner that mimics, in miniature, complex structures in the Transverse Ranges of southern California, so Rainbow Basin and Owl Canyon are university favorites for teaching geologic field mapping techniques. For these reasons, and many more that you may discover for yourself, Rainbow Basin was declared a National Natural Landmark in 1972.

Paleontologists found fossil bones in the Mud Hills in the first two decades of the twentieth century and have since combed the area so thoroughly that few remain on the surface. But fossils do weather out of the rock, and every hard rain reveals a few more. If you find

railroad facilities. Barstow was founded in 1886 as a railroad town, and for many decades the railroad was its main employer. In fact, Barstow was named for William Barstow Strong, then president of the Santa Fe Railroad. His middle name was chosen because his last name had already been used for a Santa Fe station in Kansas.

One mile from the last curve, turn left (north) onto Irwin Road (to CA 58). This turn is 400 feet beyond the 50-foot-high, red-brown volcanic knob (34°54.423N, 117°01.331W), which is one of several Miocene dacite knobs and lava flows in the Barstow area. Proceed 0.6 mile north on Irwin Road, passing through two stop signs (after stopping, of course) at CA 58. Travelers coming to Barstow from Mojave or Bakersfield via CA 58 join the route at the first stop sign. Five and a half miles north from the second stop sign, bear left onto Fossil Bed Road, a wide, well-graded gravel road marked by a BLM sign on the right labeled "Rainbow Basin Natural Area" and "Owl Canyon Campground." Drive 2.8 miles and then turn right to the entrance to Rainbow Basin.

any fossils, please leave them in place and notify the Bureau of Land Management's Barstow Field Office at (760) 252-6000. It is important that they pinpoint exactly what layer the fossil is in. Rock and fossil collecting is prohibited within the National Natural Landmark.

The many canyons and twisted arroyos in the Mud Hills contain a treasure trove of fossils of vertebrate animals that roamed the landscape about 15 million years ago. The fossil remains are remarkably diverse and include bones, teeth, and tracks of primitive bears, dogs, antelopes, rhinos, pigs, horses, camels, mastodons, rodents, and extinct pig-like animals called oreodonts. Insect and plant remains, including oak, poison oak, palm, and juniper, are also present. The complete assemblage of fossils suggests that the climate was similar to that of coastal southern California today, probably with 15 to 20 inches of rainfall per year. John C. Merriam, the paleontologist who first studied this area and named the Barstow Formation, also concluded from abundant and diverse fossil remains that the climate was warm at that time with alternating dry and wet seasons. The area was a plain, much like African savannas, where large herds grazed on lush growth of grasses and foliage. Many of the fossils have been assembled in the Raymond M. Alf Museum of Paleontology, which is part of the Webb Schools, Claremont, California, and is the only nationally accredited museum of paleontology on a secondary school campus in the United States.

The Rainbow Basin fossils are found in the Barstow Formation, which consists predominantly of mudstone, siltstone, and sandstone deposited in a shallow lake. The several layers of bright-white rock interlayered with these sedimentary rocks are rhyolite tuff composed of pebbly volcanic ash. Isotopic dating of the ashes, utilizing the decay of potassium to argon, indicates that they are between 19 and 13 million years old, so that must also be the age of the Barstow Formation. The formation was deposited long after dinosaurs vanished 65 million years ago and long before the first hints of humanity appeared in Africa, perhaps 3 million years ago.

This vignette is divided into two parts: The first part is a 4.5-mile-long drive through Rainbow Basin, where weathering and erosion have exposed a breathtaking cross section of the basin that is both scientifically spectacular and visually exhilarating. It is geologic eye candy at its finest. The second part is an approximately half-day hike up Owl Canyon; if you are up to it, this adventure includes some scrambling over boulders and up dry waterfalls. Spectacular in its own right, it is an opportunity to see, touch, smell, and taste the sedimentary strata of Mud Hills that are so well exposed in Rainbow Basin.

Drive through Rainbow Basin

You can drive a one-way round-trip loop road through Rainbow Basin. As you head north from Fossil Bed Road, you will pass the turnoff to Owl Canyon Campground on your right. The Rainbow Basin road shortly proceeds through a tight little canyon that floods during storms. Notice that the rock layers (ahead) dip to the north, away from you. After several tight twists and turns, the road curves west into an open amphitheater.

Notice that all the rocks ahead of you here dip south, toward you. The reason for this soon becomes clear. Drive west a few hundred yards to where the road climbs a small hill to a parking area on the north, or right, side (35°01.87N, 117°02.00W). Walk east about 100 feet to a saddle between two knobs for a good view of the upper beds of the Barstow Formation.

Observe that all rocks on the north wall of the basin dip south and all those on the south wall, through which you just came, dip north. The two walls are on opposite flanks of a broad synclinal trough, which is clearly visible in the east wall of the basin. This is the Barstow syncline, a fold so obvious and beautifully displayed that it is a geological celebrity; its photograph is featured in many textbooks.

The Barstow syncline, looking east from the parking area, in Rainbow Basin. The thin layer of pink, flat-lying alluvium on the skyline ridge was deposited on an eroded surface of the tilted Barstow Formation beds, forming an angular unconformity. The flat surface in the middle foreground is a low-lying version of this same unconformity.

The Barstow syncline is like a wrinkle in a rug, formed as crustal forces compressed the sedimentary layers. The trough of the fold trends from east to west, so the compressive forces must have been applied from the north and south. The Mud Hills lie athwart the Calico fault, a major structure that reaches from the eastern San Bernardino Mountains northwest across the Mojave Desert almost to El Paso Mountains. The Calico fault has the same strike and slips the same way as the notorious San Andreas fault—as do a swarm of others nearby, including the Pisgah, Helendale, and Camp Rock faults. The Camp Rock fault broke in the June 1992 M7.2 Landers earthquake. Slip on each fault is horizontal and right-lateral, such that if you stand on either side of the fault, you see the opposite side moving to your right. We'll see a small version of such a fault in Owl Canyon.

The Calico fault bends in a few places, notably in the Mud Hills. When rocks on opposite sides of a fault press into each other at the bend, they buckle into folds. The Barstow syncline formed because the Calico fault bends to the left. The situation resembles the bend in the San Andreas fault. South of Palm Springs, the San Andreas fault strikes about 50 degrees west of north. North of that resort town, the fault assumes a more westerly strike (about 60 degrees west of north). Near the southwest end of the San Joaquin Valley, north of Los Angeles, the fault again strikes more northerly (about 40 degrees west of north). The fault segment between Palm Springs and the San Joaquin Valley is known as the Big Bend.

Several branches of the Calico fault are visible from the parking area. One cuts the east wall of the basin about 500 feet north of

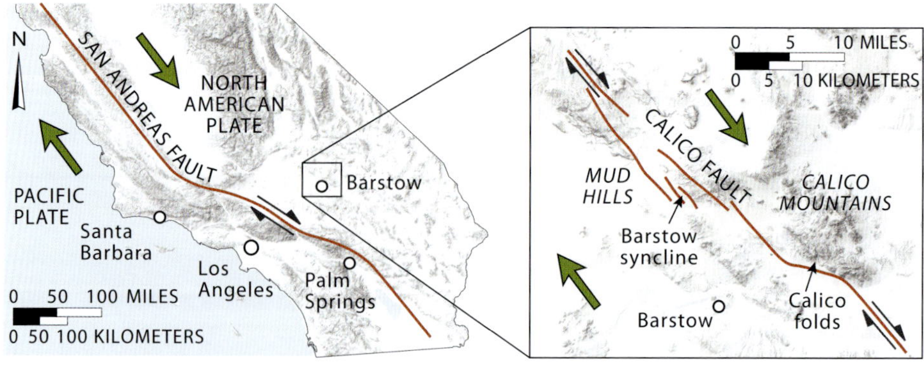

Folds, thrust faults, and earthquakes happen at bends in strike-slip faults. North of Los Angeles, a left bend in the San Andreas fault raised the Transverse Ranges. North of Barstow, a left bend in the Calico fault system produced spectacular folds in the Mud Hills and Calico Mountains.

SAN ANDREAS FAULT

All of California west of the San Andreas fault is part of the Pacific Plate, moving northwest relative to the North American Plate, which contains northern and eastern California and the contiguous United States. Give the San Andreas fault another 10 or 20 million years, and coastal southern California will slide alongside San Francisco, which lies east of the fault—thus providing rich food for political and cultural thought and making San Francisco an inland city. As the Pacific Plate and its passengers move northwestward, parts of California near the Big Bend, especially the east-west Transverse Ranges, rise because the plates push into each other. The results of that encounter include mountains, folds, and earthquakes, such as the July 1952 M7.5 Kern County earthquake, the February 1971 M6.7 San Fernando earthquake, and the January 1994 M6.7 Northridge earthquake, each of which was triggered by crustal shortening around the Big Bend.

the synclinal trough. This fault, the Owl Canyon branch of the Calico fault, is easy to spot because it displaces sedimentary layers in the syncline's south-dipping limb and places them against layers of a different color. Turn around and look at the west wall of the basin, and you will see one or two other faults. A close look will reveal that many small faults cut the rocks throughout the basin. The faults strike northwest and have a few tens of feet of displacement. These little faults are parts of the mechanism by which the Mud Hills deform.

If you look at the east wall of the basin from the parking area, you can see an interesting geologic structure—an angular unconformity—between the tilted, layered rocks and a thin cap of pink sand and gravel with Joshua trees growing on it. Geologists adore angular unconformities because they record so much geologic history, especially tectonic events. This one records folding of the lower, 15-million-year-old, layered rocks, their uplift and truncation by erosion, deposition of the sand and gravel above, then more uplift and erosion. The sand and gravel cap is no more than 1 million years old, so the buried erosional surface of the unconformity represents about 12 million years for which no geologic record exists at this place.

Several volcanic ash beds are visible from the parking area. Most conspicuous is the Skyline Tuff, a white to buff layer about 10 feet thick on the narrow skyline ridge northwest of the parking area. When volcanic ash becomes consolidated to form rock, it is known as tuff (not to be confused with tufa, which is calcium carbonate deposited from spring or stream water). The tuff bed dips south, under your

View looking north from the parking area at tilted beds on the north side of Rainbow Basin. The Skyline Tuff is the prominent tan layer on the skyline with an isotopic age of about 15.0 million years—the time it erupted from a volcano. If you look at the tuff from this spot, the layers dip toward you, pass beneath you, and reemerge on the south side of the syncline behind you.

Here, on the north limb of the Barstow syncline in the north side of Rainbow Basin, the Skyline Tuff and surrounding beds are folded into a broad secondary syncline and cut by several northeast-striking faults. One fault (dashed line) dropped a piece of the Skyline Tuff down to the right center of the photo. (35°02.0N, 117°02.15W)

feet, and emerges again on the south limb of the syncline where it dips northward and is harder to spot from here. As you drive west from the parking area, the road passes close to the Skyline Tuff just before swinging sharply south. Here the Skyline Tuff on the north limb of the syncline, a few yards north of the road, is folded into a much smaller syncline. Another prominent white tuff bed is exposed in a bottleneck passage after the road exits the basin. Mud-crack impressions on the tuff's lower surface indicate that the lake in the basin dried up occasionally.

Many of the rocks in the Barstow Formation are bright green. Novice geology students tend to assume that green rocks indicate the presence of copper, a potent rock dye (pigment). However, most of the green in these rocks comes from clay minerals that form during weathering. Volcanic glasses readily alter to slippery green and brown clay.

Owl Canyon

The walk up Owl Canyon will take you downsection stratigraphically, that is, back in time, from the younger mudstone, sandstone, siltstone, and tuff of the upper Barstow Formation, which were deposited in quiet lake water, into the older breccia of the Mud Hills Formation, and finally into volcanic rocks of the underlying Pickhandle Formation. Along the way you will encounter faults, unconformities, and changes in dip, but the overall adventure will be a journey across strata laid down in this inland basin.

The layered rocks of Owl Canyon record a time when mountains rose along a fault, and as they rose, weathering and erosion worked hand in hand with gravity to keep the mountains from becoming too high. Debris shed from the mountains washed down into a basin where it was deposited as distinct layers of clay, silt, sand, and gravel that we'll see on the hike.

The trailhead is at the north end of the Owl Canyon Campground (35°01.49N, 117°01.34W). Walk to the bottom of the wash and follow it upstream, avoiding any trails leading out of the wash. Bring a flashlight. Numbered stops in parentheses in the following text are keyed to the location map of Owl Canyon on the following page.

A word of caution before you start. It is always best to let someone know where you are going and when you plan to return. Cell phones rarely work in this region, and it is smart to have a hiking partner to help in an emergency. Dehydration is always a possibility, especially in the warmer months. Bring lots of water, and watch for rattlesnakes (rarely seen but they are around).

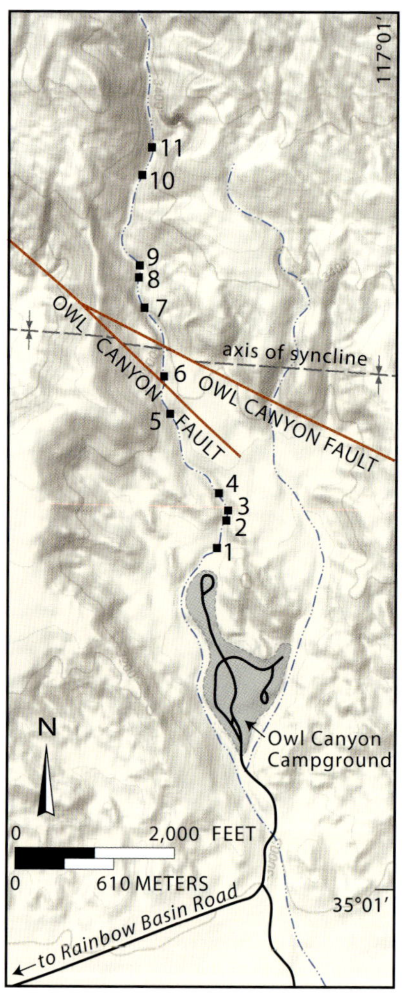

Map of the lower part of Owl Canyon hike. Numbers refer to locations of features discussed in the text.

AGE	ROCK UNIT	THICKNESS	DESCRIPTION
QUATERNARY	older alluvium (1 million years to present)	0 to 50 feet	sand and gravel
EARLY MIOCENE	Barstow Formation (19 to 13 million years)	2,150 feet	conglomerate, sandstone, mudstone, siltstone, and tuff
EARLY MIOCENE	Mud Hills Formation	760 feet	sedimentary breccia, sandstone
EARLY MIOCENE / LATE OLIGOCENE	Pickhandle Formation (24 to 21 million years)	1,780 feet	volcanic breccia and mudflows
JURASSIC	165-million-year-old granitic basement		granite

Generalized stratigraphy of rock formations in the Mud Hills, Mojave Desert. Wiggly lines indicate unconformities.

View looking west from the Owl Canyon Campground at the north-dipping Barstow Formation beds overlain unconformably by a thin pink layer of alluvial sand and gravel on the skyline ridge. Rainbow Basin is on the other side of the ridge.

Before starting your hike, look at the west wall of the canyon. Older pink alluvial gravels lie with angular unconformity upon the Barstow Formation beds. This is the same angular unconformity seen near the skyline in Rainbow Basin, and you will see it many more times in Owl Canyon. Notice that the layers of the Barstow Formation dip to the north, upstream, just as they do in the south part of Rainbow Basin. This means that you are on the south side of the Barstow syncline, and as you walk up the canyon, you will proceed through younger and younger strata until you reach the syncline's axis. Rocks there are nearly flat lying and consist mostly of mudstone, siltstone, and sandstone with occasional beds of pale white tuff. All were deposited in lakes or by high water spilling over the banks of broad streams.

About 100 yards from the trailhead, at a sharp zigzag bend (Stop 1), horizontal layers of young gravel are plastered onto the canyon wall at about eye height. They are just a coating left over from a former gravel filling of the canyon, which has been filled and sluiced countless times in its history. About 200 feet farther, the stream makes a second zigzag (Stop 2). On the east wall here, you will find a splendid exposure of an unconformity, where pink gravel is piled up against the layered Barstow Formation.

Dashed line marks the unconformity between flat-lying layers of stream gravel that were deposited upon older, tilted Barstow Formation beds at Stop 1 in Owl Canyon. Staff is 4.5 feet long. (35°01.584N, 117°01.322W)

Unconformity between geologically young, horizontally bedded pink gravels (left and center) and older, tilted Barstow Formation beds (right) at Stop 2 in Owl Canyon. (35°01.60N, 117°01.30W)

Another 100 feet or so upstream (Stop 3), again on the east side, a little hillock of green beds in the Barstow Formation is almost completely covered by younger pink gravels. Nearby, a large (4-foot-diameter) boulder of tuff breccia has been washed downstream from the underlying Mud Hills Formation, which crops out upstream. We will see more of this formation later. About 50 feet beyond the boulder, a side trail takes off up the east bank. Keep to the gully floor.

Big boulder of tuff breccia derived from the Mud Hills Formation upstream. Staff is 4.5 feet long. (35°01.605N, 117°01.307W)

About 200 feet beyond the side trail, a hard, white bed, about 1 foot thick, crops out on both sides of the wash (Stop 4). This tuff layer, like all the beds you have walked through so far, dips to the north—upstream; thus, you have been walking upsection into younger and younger rocks. The tuff beds on each side of the wash, once continuous, have been cut and displaced by a fault that strikes up the canyon. The fault itself is not exposed in the floor of the canyon, but you can determine where it ought to be by standing on the end of one of the tuff beds and noticing that the other is about 20 feet to your right. The tuff bed has been separated by the fault (which we assume to be steeply dipping), most likely by right-lateral strike-slip, although the same separation could be accomplished by purely

View looking upstream of a steep, north-dipping, white tuff bed displaced by a fault (dashed line) at Stop 4 in Owl Canyon. The fault is not exposed in the canyon floor. Staff is 4.5 feet long. (35°01.634N, 117°01.332W)

vertical displacement on a steep fault. Quiz: If displacement were vertical, which side went up? Erosion along the fault is probably why the canyon is so straight here.

A few hundred feet farther up the canyon, the rock layering flattens, and the dip begins to shift back to the south, down canyon (Stop 5). This change is complicated by several small faults that cut through the canyon at this point. Although not wholly obvious, this spot marks the trough of the Barstow syncline (35°01.74N, 117°01.40W), so that as you walk north from here, you pass downsection into older and older rocks.

Shortly after you cross the axis of the syncline, you come to what appears to be a major fork in the canyon (Stop 6; 35°01.798N,

117°01.434W). The right fork is the easier route, but either will work because this is not really a fork. The left "fork" cuts off a meander loop in the channel, and the two channels join a short distance upstream. Note how thin-bedded, gray mudstone beds dip down canyon, clearly indicating you are on the north limb of the syncline.

Within a short distance, the canyon wall becomes higher and more dissected. After a few hundred feet between close valley walls, you reach a section where the stream bank collapsed (Stop 7), making the channel wider. A little farther, large boulders of sandstone from the Barstow Formation have tumbled into the canyon, blocking the channel (Stop 8).

Tumbled boulders are the effects of weathering, erosion, and landsliding in a sandstone member of the Barstow Formation where Owl Canyon widens at Stop 8. Arrow points to 4.5 feet long staff. (35°01.959N, 117°01.487W)

Around a bend about 150 feet beyond the fallen blocks, look for the entrance to a natural tunnel in the east wall of the canyon (Stop 9). You will need your flashlight to explore it. About 20 yards long, this tunnel links Owl Canyon to a small tributary canyon on the other side of the ridge. During heavy rains, water flows from the tributary canyon through the tunnel into Owl Canyon. Natural mud tunnels are quite unusual, but they are fairly common in Rainbow Basin. This is one of the best in the area.

Entrance to an unusual tunnel (lower left) on the east side of Owl Canyon (Stop 9). Bring a flashlight to explore it. (35°01.97N, 117°01.47W)

Tunnels like these are created when water seeps through a natural crack or fracture in relatively soft or poorly consolidated sediment. With time, erosion widens the crack to the point where it becomes, in this case, a tunnel. Known as piping, this phenomenon caused the Teton Dam in Idaho to fail catastrophically in June 1976 and perhaps also the failure of St. Francis Dam in 1928 (Vignette 21). Such tunnels are relatively harmless in an undeveloped desert environment, but deadly if you are caught in a flash flood or sudden collapse of the tunnel.

After a major bend a few hundred feet beyond the tunnel, the wash narrows impressively, with steep walls closing in on either side. Eventually, below the base of the Barstow Formation, you will come to coarse breccia beds of the Mud Hills Formation. Examine the cobbles in the breccia. They are angular fragments of granite, a salt-and-pepper textured rock made of large (0.25 inch or so) crystals of quartz, feldspar, and mica. These cobbles are identical to the granite exposed in the north part of the Mud Hills. Clearly, the breccia records a depositional environment quite different from a quiet lake in which mud was deposited, to a more active alluvial fan environment in which rock avalanches and torrential rains brought boulder-laden landslides and debris flows into the basin.

After about 500 feet of this narrow section, a fault crosses the wash bottom (Stop 10). The fault dips about 80 degrees south (downstream) and has brought finer, less resistant sandstone and mudstone on the north into contact with more resistant, coarse pebbly and cobbly sandstone on the south. About 70 feet upstream of the fault is a huge (8-foot-diameter) boulder of brown Pickhandle Formation breccia containing sharply angular granite fragments. Some of the pieces are more than 1 foot across. Feel free to pat the boulder or sit upon it to rest but be sure to say please and thank you.

A few hundred feet farther upstream, the wash walls become high and vertical and seem to hang over you. Near the top they are full of small holes and pits, called tafoni, caused by differential weathering of the rock.

The easy walk ends at a dry waterfall (Stop 11) carved out of the more resistant rocks of the Pickhandle Formation. You have now walked completely through the Barstow and Mud Hills Formations. The layers in the Pickhandle Formation here dip about 20 degrees steeper than those in the overlying Mud Hills Formation, so the contact between the formations is another angular unconformity (but it is not well exposed).

From this point north, the canyon walls become much steeper and the rocks consist of much coarser fragments. Gone are the smooth, rounded pebbles and cobbles of the conglomerate. The Pickhandle Formation is a stack of volcanic breccia (angular gravel) and mudflows

Boulder of Pickhandle Formation breccia on the floor of Owl Canyon, about 200 yards downstream from the dry waterfall. Staff is 4.5 feet long. (35°02.075N, 117°01.503W)

A dry waterfall in the narrow, upper part of Owl Canyon is cut stepwise into the Pickhandle Formation volcanic breccia. The canyon at Stop 11 is about 8 feet wide. Staff is 4.5 feet long. (35°02.146N, 117°01.469W)

deposited 24 to 21 million years ago. Where the canyon is narrow and impressively deep, the walls are made of large, shattered blocks of granite interbedded with sedimentary layers. These blocks slid off fault scarps and into what would eventually become the Pickhandle Formation. The deposit is a smaller version of a rock avalanche landslide, such as the much larger Blackhawk Slide (Vignette 16), and tells a geologic story of robust faulting and mountain building in this area.

Beyond the dry waterfall and a little farther north of it, you will find that Owl Canyon terminates on the slopes of a granite ridge that rises to the crest of the Mud Hills. These granite slopes with their Joshua trees and rounded outcrops are the source of the granite boulders and cobbles in the Pickhandle and Mud Hills Formations. This is an excellent place to stop for lunch and ponder the geologic history of this colorful area and the enormous amount of geologic work that went into making the Mud Hills what they are today.

VIGNETTE 14

AMBOY AND PISGAH CRATERS

Young Volcanoes in the Mojave Desert

SAN BERNARDINO COUNTY

How and why volcanoes erupt exactly when and where they do is not always clear, although a few can be tied directly to a fault or other geologic structure that might localize the eruption. For example, subduction-related volcanoes, such as those in the Cascades of Oregon and Washington, occupy specific, predictable positions relative to the continental margin. But a subduction zone does not lie beneath the thirty or so relatively young volcanoes that dot the triangular region of the Mojave Desert between Barstow, Baker, and Twentynine Palms. Amboy and Pisgah Craters are among the youngest and most easily explored of those volcanoes. Both have small craters at their summits, but mainly they are excellent examples of cinder cones. Let us consider a few things about cinder cones before we begin our explorations of them.

Cinder cones are small basaltic volcanoes consisting of frothy basaltic lava fragments an inch or two in diameter, the sort of rock you might buy in bags to spread beneath your shrubs or to line the bottom of your barbeque. Geologists call such material *tephra*, a term Aristotle used based on the Greek word for "ashes." Tephra refers to any fragmental material (ash, cinders, or even refrigerator-sized blocks) that a volcano blasts into the air. Volcanic bombs are viscous globs of molten lava erupted, shaped, and cooled while flying through the air.

Cinder cones typically form and erupt tephra for a few weeks or years, produce lava flows, and then go quiet, usually forever. In most cases, the next eruption in the area will build a new cinder cone. Initially, coarse tephra erupts around the vent to build a cone; finer tephra drifts downwind as an ash cloud. Most cinder cones are symmetrical, although a strong wind may shape them. Concurrent with the tephra eruption, the volcano may produce a lava flow. The flows commonly extrude from the base of the cone because the pile of loose cinders is too weak to support an internal column of magma. In many cases, a segment of the cone is carried away on a lava flow, producing

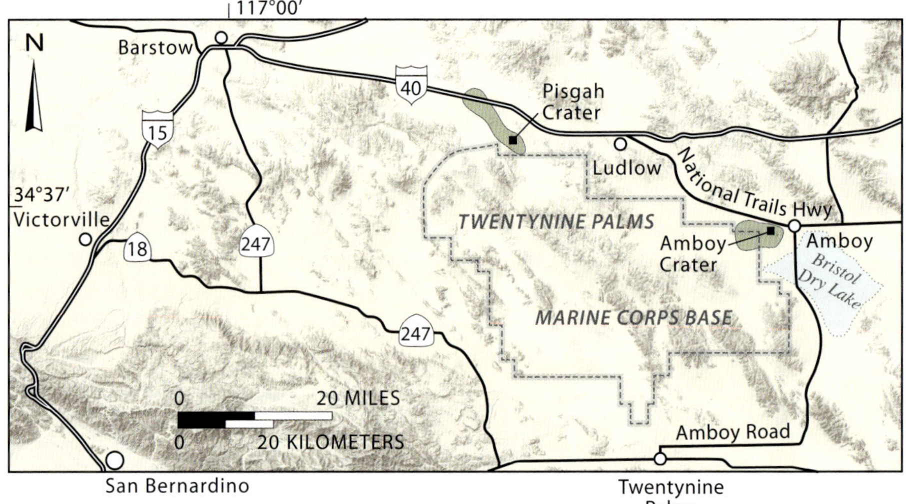

GETTING THERE

Amboy Crater is easy to reach from Barstow or Twentynine Palms. From Barstow, drive 52 miles east on I-40 to Ludlow. Turn south (right) on the off-ramp, then east (left) at the stop sign. Drive 26 miles east on National Trails Highway (old US 66) to a sign marking the entrance to Amboy Crater. This turnoff is 1.3 miles east of the bridge where the highway first crosses onto lava. It is 1 mile west of the intersection of Amboy Road and the National Trails Highway. If you're coming from Twentynine Palms, take Amboy Road east and north 49 miles. You will cross onto the playa of Bristol Dry Lake about 9 miles south of Amboy. Continue to the National Trails Highway and turn west (left). Drive 1 mile west to a turnoff on the left at the sign for Amboy Crater. Drive into the parking area (34°33.43N, 115°46.87W), and then you can wander short distances into the lava field south or west of the parking area. The trail to the cinder cone is marked and commences on the west side of the parking area.

Pisgah Crater is 40 miles east of Barstow and 2 miles south of I-40. See the directions with the map for Pisgah Crater later in this chapter.

Examples of tephra. Spindle-shaped bomb (left), angular block (top), red-brown cinder (right center). Along the bottom are spindle-shaped bombinos, a form of tephra blasted out of a volcano as molten droplets of lava and twisted in air like a football tossed in an arc-shaped ballistic trajectory. Look for bombinos on the northeast flanks of Pisgah and Amboy Craters.

a breach in the cone wall. The west side of Amboy and the southeast side of Pisgah have just such breaches in their cones.

Amboy and Pisgah Craters stand amid large fields of basaltic lava and are easily seen from the highway. Although their tephra eruptions undoubtedly made good fireworks displays, they were neither dangerous nor violent by volcanic standards. You could have watched them erupt safely from as little as 1 mile away. Contrast that to rhyolite eruptions, such as that of the Bishop Tuff from the Long Valley caldera, whose ash covered half of California 765,000 years ago. Such a cataclysmic eruption can devastate a region the size of an average state and change the world's climate for years.

Most of the basaltic lava flows that erupted from Amboy and Pisgah Craters have pahoehoe surfaces. *Pahoehoe*, a Hawai'ian term, refers to surfaces that are relatively smooth and covered in places with distinctive ropy coils. This ropy texture forms when molten lava

PARÍCUTIN CINDER CONE

The world's most famous cinder cone may be Parícutin, 200 miles west of Mexico City. It unexpectedly appeared in a cornfield in 1943 and then grew to a height of 1,200 feet during the next 9 years, erupting lava flows that covered more than 10 square miles and buried a village. It is a giant among cinder cones. Although the sudden appearance of Parícutin certainly startled the farmer who watched the eruption begin in his cornfield, it should not have been much of a surprise. The surrounding area has about 200 cinder cones.

Broken and tilted slab of pahoehoe lava with a characteristic ropy surface. Hammer handle is 14 inches long. (34°44.82N, 116°24.77W)

rumples its thin outer skin of hardening rock, much like the surface texture of fudge that cools as it is poured into a pan. Some of the flows have *a'a* surfaces, another Hawai'ian term. A'a surfaces are rough, clinkery, and jagged; they can tear your clothing and destroy your boots.

Although they look quite different, flows with pahoehoe and a'a surfaces solidify from the same kind of magma; they have the same chemical and mineral composition. The difference in flow surfaces depends upon how much gas the molten lava contains and the speed of the flow. Volcanic gas, the principal component of which is steam,

Smooth pahoehoe flowed over rough, clinkery a'a at Pisgah Crater. Orange knapsack is about 18 inches tall. (34°44.53N, 116°22.56W)

makes basaltic lava more fluid and therefore more likely to develop a pahoehoe surface; degassed lava is considerably more viscous and tends to develop a'a surfaces. Indeed, both surface textures may form on the same flow; a smooth pahoehoe surface near the vent may transition to a rough a'a surface farther away, reflecting the flow's loss of gas to the air and increasing viscosity.

How long ago did Amboy and Pisgah volcanoes erupt? They are too young to date by the potassium-argon technique, which can be used on basalt erupted as recently as about 100,000 years ago. They may be young enough to date by the radiocarbon (carbon-14) technique, which can be used to date samples as old as about 50,000 years, but this requires carbon-based material to date, normally volcanically charred wood. So far, no one has found any wood around or beneath either cone. Neither volcano has much evidence of erosion, but that is typical of cinder cones; the loose tephra absorbs water so well that surface runoff rarely makes gullies. The best age estimates, from various methods, are about 25,000 years for Pisgah and about 80,000 years for Amboy.

At present, no one can predict exactly when or where another eruption might occur. An eruption in an unpopulated area would be a spectacular fireworks show; an eruption in a populated area or beneath a hazardous waste dump would be a disaster. Imagine the scene if a volcano like Pisgah or Amboy were to erupt near Las Vegas or Lake Mead! The highways would be clogged with geo-tourists coming from all over the Southwest to see the show. For this and scientific reasons, the young volcanoes of western North America will remain under close scrutiny.

The best way to enjoy these youthful craters is to get them underfoot. Be sure to bring plenty of water; black basalt really gets hot in the sun. Hiking is easy, except across a'a surfaces or deep fissures. The fantastical lava shapes, the interplay among lava, wispy wind-driven sand, and the creatures that live in the rocks and leave their tracks on the sand are all worth seeing. Keep a wary eye open for rattlesnakes; they also live among the rocks and leave their tracks on the sand.

Amboy Crater

You can hike over lava flows up and into an absolutely undisturbed cinder cone at Amboy Crater, a National Natural Landmark. It is 3 miles west-southwest of Amboy, which was a lively settlement in the days when US 66 was a major transcontinental artery. (Amboy, by the way, marks the west end of an alphabetically named series of watering stops along the Santa Fe Railroad—Amboy, Bolo/Bristol, Cadiz, Danby, Essex, Fenner, and so on.)

The 220-foot-high cone is about 1 mile from the parking area, and its circumference is another mile, so the round-trip hike is roughly 3 miles. Allow three hours, minimum. The whole trek is a dandy adventure, a miniature expedition. Remember to carry plenty of water. Follow the trail with its occasional marker posts from the parking area to the west side of the cone. An easy walk brings you to the base of Amboy Crater in less than 30 minutes.

Most of the lava you see along the trail is pahoehoe. The flow surface is full of gas pockets and bubbles left where volcanic gas did not quite escape. Although the top of the flow is relatively smooth overall, much of it is not really a very good example of pahoehoe because it lacks the typical ropy surface.

While walking on these flows, notice the number of closed depressions on the flow, some elongate, others irregular, most about 10 feet deep. Rubble and sand cover their floors. They indent a relatively flat lava surface, and flows tilt down around their edges, so that they appear to be collapse features. They probably formed when molten lava drained out from beneath a solidifying crust.

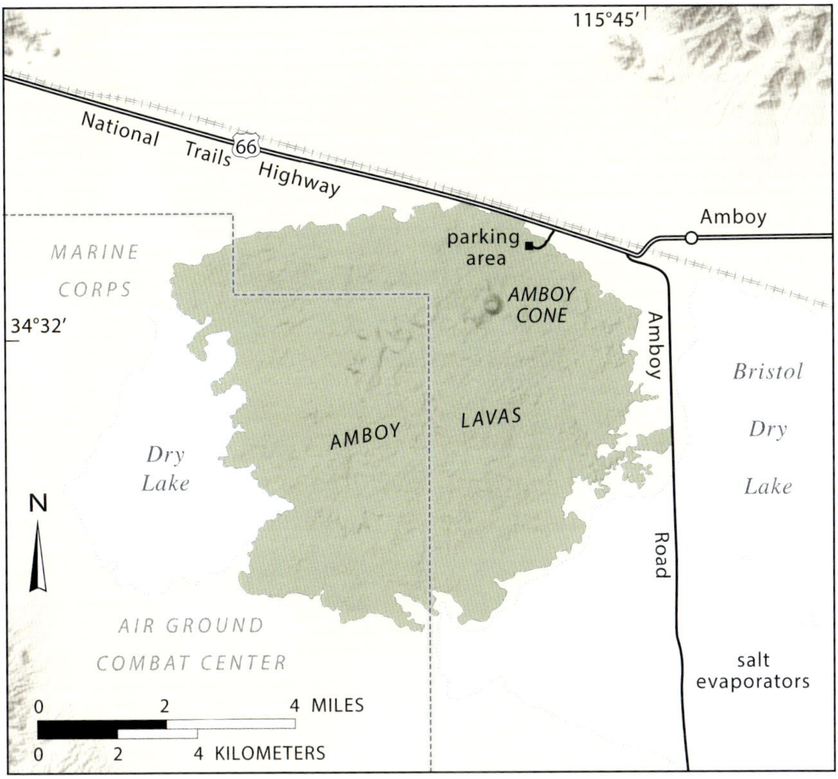

Location and access map for Amboy Crater.

Aerial view looking east of Amboy Crater surrounded by basaltic lava flows blanketed with wind-blown sand. Lava flows on the downwind side of the cone are barren of sand. (34°32.7N, 115°47.43W)

Many of the lava surfaces are cracked into four-, five-, or six-sided polygons a foot or so across. These are the tops of vertical cracks that form when the lava shrinks slightly while it solidifies but while the lava is still very hot. Similar cracks form the palisades of vertical columns you may see in basalt flows exposed in other special places like Devils Postpile National Monument in the Sierra Nevada and the Giants Causeway in Ireland, where five-, six-, and seven-sided columns are most common.

SALT MINING

Bristol Dry Lake, the great expanse of flat ground you can see southeast of Amboy Crater, is a desiccated remnant of a much larger lake that flooded this undrained desert valley about 20,000 years ago when the climate was much wetter than now. It contains salts, chiefly sodium and calcium chlorides, which weather out of rocks in the surrounding mountains. They accumulate on the desert floor because the basin has no outlet stream, so the salts become concentrated by evaporation of whatever water collects there. A sort of mining operation harvests the dissolved salts through a system of ditches that drain brine from the sediments in the valley floor and carry it to large evaporating flats. Look for white salt crusts on the sides of the ditches.

Saline pool in an excavated ditch adjacent to Amboy Road, Bristol Dry Lake, 5 miles south of Amboy. (34°28.24N, 115°44.56W)

Sand-filled depression within a pahoehoe lava flow. (34°33.12N, 115°47.11W)

Polygonally cracked surface of a tilted pahoehoe lava flow. Staff is 4.5 feet long. (34°33.430N, 115°46.934W)

As you explore the lava flows on your way to the cinder cone, look for pressure ridges—great slabs of hardened lava that buckled up due to pressure from still-liquid lava flowing beneath the congealing crust. These linear ridges, with open cracks along their crests, form most commonly in pahoehoe flows, especially those associated with Pisgah Crater. Look for places where molten lava squeezed up through the crack in the crest of the pressure ridge to make great blobs that resemble black taffy. These bulbous extrusions are called squeeze-ups.

Breached pressure ridge in pahoehoe lava. The white material in the bottom of the crack is sand. The hammer handle (circled) is 14 inches long. (34°44.970N, 116°24.788W)

Instead of climbing 220 feet up the steep trail on the north side of Amboy Crater, follow the trail counterclockwise around the base to the breach in the west wall, where the climb is only 80 feet to the floor of the inner crater. A trail inside the crater ascends a gentle slope to its rim, which you may circle around for good panoramic views of the entire 24-square-mile lava field and the surrounding terrain. The loose rubble on the cone makes it hard to descend any slope without slipping. Be careful.

Several phases of the eruption history can be read from the shape of the cone. The main edifice formed first during the main period of tephra eruption. A lava flow then breached the west side of the cone. You can see a bit of that flow within the breach. Two minor explosive pulses inside the main crater produced two small nested tephra rings on the crater floor.

Leave the crater through the breach in its west wall and walk cross-country and counterclockwise around the cone's base to the south side of the cone to a large depression with a flat floor about 40 feet below the level of the lava flow surface. It is best viewed from a small bench at the base of the cone. Slopes above the bench display scars of a landslide, which probably created the bench. You can reach the floor of the flat by way of one of the gullies in the bench.

Along the way, you can see why Amboy Crater has escaped cinder mining. Much of the cone consists of solid lava chunks, 20 inches or so in diameter, and of even larger masses of lava blobs that stuck together while still partly molten. Commercial operators prefer very frothy cinders in the pea to walnut size range. Such material is scarce in Amboy Crater, a deficiency to which Amboy owes its preservation, in contrast to Pisgah Crater, as we will learn below.

Upon crossing the flat floor of the depression at the south base of the cone, you will come to a lava flow of shiny, black basalt with a convoluted pahoehoe surface. Look for a band of lava close to the base of the cone that is free of rubble, somewhat smoothed and worn, and a little lighter in color. A gully more than 3 feet deep runs from the base of this band across the flat floor of the depression. Its appearance suggests that quite a lot of water flows occasionally down the face of this band. It is a dry cascade. If you step up the 20 feet of blocky wall to its top, you will see a dry streambed of clean rock curving out of sight around the east side of the cone. Where did this water come from?

Almost opposite the dry cascade, the slopes of Amboy Crater change from large chunks of lava to small rectangular chunks of surprisingly uniform size, between 2 and 5 inches across, along with a goodly sprinkling of spindle-shaped volcanic bombs and bombinos. In contrast to the outer surface of the cone's southern slope, smooth, deep rills and gullies dissect the cone's eastern slope, suggesting it may be an older, more eroded part of Amboy Crater. Alternatively, the east side of the cone may be gullied because it is on the downwind side. Silt blown over the cone may be trapped there among the cinders. This silt would then prevent water from passing through the cone, causing it instead to course down the cone's outer surface and carve gullies.

In another 200 yards or so, still at the eastern base of the cone, you come to a larger and deeper gully with tributaries. The streambed ends here. Look up the slope to see several pipe-like openings about 4 feet in diameter on the walls and at the heads of the gullies. These openings look like pipes coming out of the cone, and indeed they are natural pipes, but their actual diameters are much smaller than the cavernous opening you see. Water flowing from the crater through these natural conduits eroded the slope gullies, flowed along the stream course, poured over the now-dry cascades, and dissected the floor of the depression. Inclined layers of small uniform blocks, which form the wall of the cone, show up nicely in the gully.

Rilled slopes, but without pipes and gullies, are on the northeast side of the cone. You can see a distinct contact between them and the smooth younger slopes. It appears that more than half of an older, rilled cone was partly destroyed before the younger cone covered most of it. The contact between the two, which dips down to the west at an angle of about 45 degrees, is more obvious if you look back from the broad flat area north of the cone.

Your car, in the parking lot, should come into sight to the north as you round the northeast corner of the cone and regain the trail.

View looking south at Amboy Crater cinder cone with deep, closely spaced rills on its northeast flank. (34°32.72N, 115°47.33W)

Pisgah Crater

People who do not want to make the long trek to Amboy Crater may wish to visit Pisgah Crater (it is closer to Barstow), but cinder quarrying has greatly disfigured the cone, making some parts of it off limits. A rest stop about 28 miles east of Barstow is situated at the westernmost extent of the Pisgah lava flows. From there you can see Pisgah Crater, the source of these lava flows, 9 miles to the east-southeast. Interstate 40 crosses lava flows for about 2 miles until you come to a sign that announces "Hector Road Exit, 1 mile." Exit I-40 at Hector Mine Road (exit 33), turn south (right) and then east (left) at the T intersection with the National Trails Highway (old US 66). Drive east 4.5 miles to Pisgah Quarry Road (34°46.31N, 116°22.44W), the haul road that provides access to the cinder cone and the lava flows around it.

Geologists have determined that Pisgah Crater and its lava flows formed during three eruptive episodes about 25,000 years ago. The main cone and a'a flows formed in the first episode. That lava is characterized by tiny plagioclase and olivine crystals that can be seen only with the aid of a hand lens.

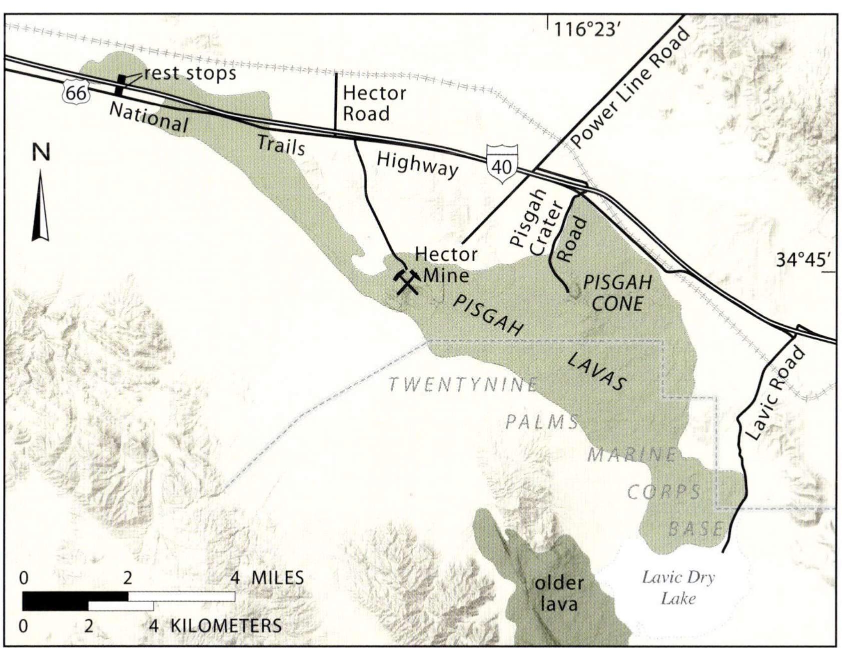

Location and access map for Pisgah Crater and associated lava flows.

Aerial view looking east at Pisgah Crater. The volcano has been extensively modified by cinder quarrying operations. The flat bench around the volcano in the foreground consists of volcanic sand and dust, the spoils from those operations.

Most of the surrounding lava field, consisting of both extensive a'a and pahoehoe flows, erupted during the second episode. This lava's olivine and plagioclase crystals are about an eighth to a sixteenth of an inch long, large enough to see with a naked eye.

The third eruptive episode added some tephra to the northeast quadrant of the cone, and a pahoehoe lava flow erupted from a vent in the floor of the main crater, burrowed its way beneath the southeast quarter of the cone, and then flowed around the south side of the crater. Its lava is distinguished by abundant, large, white plagioclase crystals up to a half inch long and sparse, tiny olivine crystals.

A series of stops along Pisgah Crater Road provides good examples of the lava flows. The first is at a wide place about 0.2 to 0.4 mile south of the National Trails Highway. Walk east 200 to 300 yards across a'a lava largely covered with windblown sand, carried by prevailing westerlies from Troy Dry Lake, which lies between Barstow and Pisgah.

Now drive about 0.6 mile south on Pisgah Crater Road from the National Trails Highway, park, and walk about 50 feet eastward to the front of another a'a flow. Sand buries only about 15 to 20 percent of this flow, which is younger than and lies upon the flows of the previous stop. You might suspect that this difference in sand cover reflects a difference in age. Rock magnetism, however, indicates that the three main sets of lava flows at Pisgah Crater, including these, erupted within a few decades or centuries of each other. The difference in cover, therefore, probably reflects a difference in sand supply. When you return to your car, look back at the abrupt front edge of the younger flows rising 7 to 10 feet higher than the flows at the first stop.

Continue driving along Pisgah Crater Road and stop where it curves west 1 mile from the starting point on the National Trails Highway. Here the road crosses a small lobe of a'a lava that flowed westward down a shallow gully (34°45.60N, 116°22.90W). Basalt lava is so fluid that it will follow even minor gullies. A basalt lava flow can be diverted with an artificial embankment, as has been done in Hawai'i and Iceland with varying degrees of success. Walk east onto the flows. At first glance, they look younger than those at the last two stops because less sand covers the surface. But the surface here is covered with platy chips of basalt, which also cover patches of sand that fill the low places. Sand is actually more abundant than appears at first glance. These lavas are about the same age as those at the preceding stop.

Continue south on Pisgah Crater Road to a parking area (34°45.18N, 116°22.94W) outside a gate 1.5 miles south of the National Trails Highway. Inside the gate and 100 feet to the east you will find a dark, rough, young-looking lava flow with very little sand cover. It has a typical clinkery a'a surface. Other nearby flows have smooth or ropy pahoehoe surfaces.

Walk east-northeastward over or around the north tip of the lava flow to a point 250 yards out of sight of the parking area and onto a dome-shaped area about the size of a football field covered not by black lava, but by rounded gray, tan, and red-brown cobbles. This little hillock, ringed by lava flows, is a window through the lava flows to a sample of the desert surface upon which Pisgah lavas flowed. Such a landform surrounded by younger lava takes the Hawai'ian name *kipuka*, meaning "hole"—thus a hole within the lava field. Notice that the cobbles are diverse kinds of varicolored volcanic rocks derived from the Cady Mountains about 5 miles north of Pisgah Crater. Their rounded shapes indicate they were transported to this site by streams and debris flows. Afterward, lava flowed from Pisgah Crater over the old desert floor but did not cover this patch.

To see the differences between lavas produced during the first and second periods of Pisgah's three episodes of eruption, return to the paved highway and drive 2 miles east on the National Trails

Highway. Park where the highway bends north (left) to cross the railroad tracks. Walk south about one-quarter mile onto sand-covered lava flows (34°45.036N, 116°20.970W) that form a broad plateau pocked by numerous lava tubes and depressions, some as much as 50 feet across. These collapse sinks formed when molten lava drained from beneath the hard crust of a lava tube or pond. Carefully chip off

View looking northwest of younger Pisgah Crater basaltic lava flows surrounding a kipuka (circled) with its rounded, stream-worn, pink and gray cobbles derived from the Cady Mountains on the skyline. (34°45.22N, 116°22.78W)

LITHIUM MINE

Extensive tailings of pale-gray tuff 1 mile west of Power Line Road are from Hector Mine, an open-pit mine that produces hectorite, a rare, lithium-bearing clay. Hectorite has a lithium content of 1.2 percent, which is a huge concentration of an element that the stars largely forgot to make. The tuff formed and altered to clay in restricted alkaline lakes through heating by hot spring activity along fault zones. The clay has a greasy feel. It is used in cosmetics and is refined to yield lithium metal for batteries and electronic devices. The nearby Hector fault produced a magnitude 7.1 earthquake on October 16, 1999.

a piece of lava and examine a fresh surface. Notice that crystals are not visible without a hand lens—a characteristic of the first eruptive episode lavas. They are overlain by extensive pahoehoe flows from the second episode. Walk a little farther south and climb carefully up onto the lava bench to look at a piece of this second flow with its abundant small crystals of white plagioclase feldspar and clear, pale-green olivine.

Good examples of ropy pahoehoe may be seen along Power Line Road, a well-maintained gravel road, exactly 1 mile west of Pisgah Crater Road. Do not confuse it with another power line road only 0.1 mile west. Power Line Road trends southwest following a triple set of mammoth power lines that carries electricity from Hoover Dam to Los Angeles. The road forks 0.2 mile from the National Trails Highway; bear left onto road NR8685 and follow it for about 2 miles to where it encounters pahoehoe lava flows.

Natural sand blasting by the wind has enhanced the somewhat linear vesicles, or gas cavities, in this basalt. Sunglasses for scale.

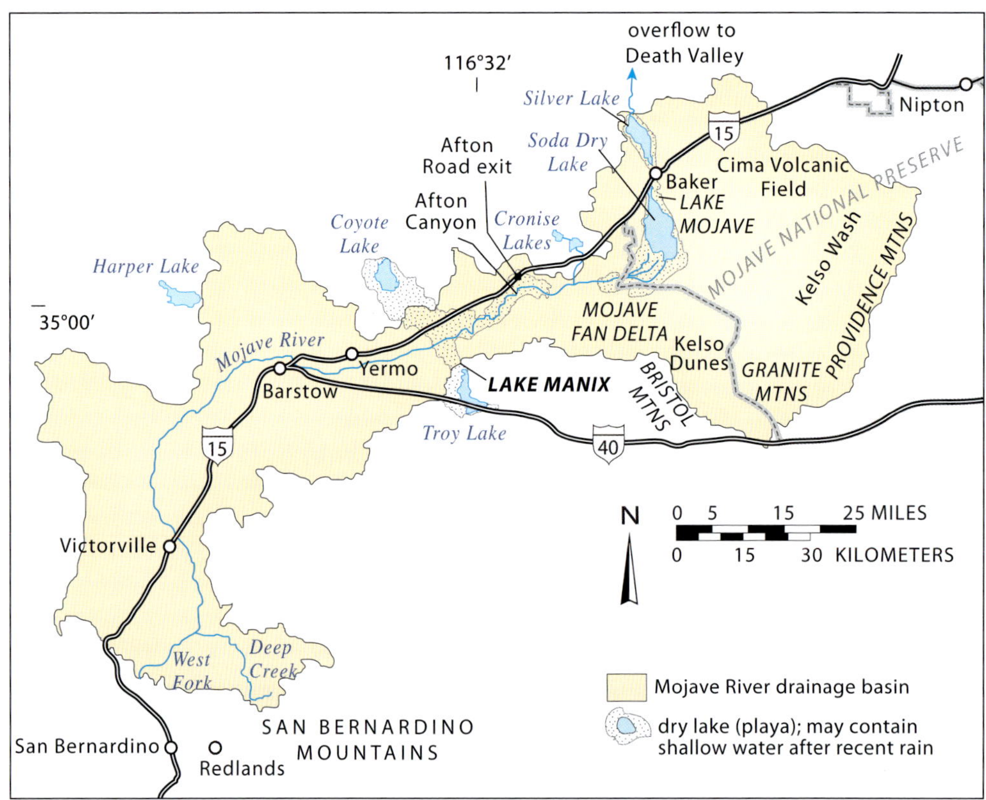

—Modified from Enzel and others, 2003

GETTING THERE

Interstate 15 follows the Mojave River from its headwaters in the San Bernardino Mountains to its modern termination at Soda Dry Lake, passing through areas occupied by Lake Manix during the last ice age. Between Victorville and Barstow, the river is west and out of sight of I-15. It turns eastward in Barstow and crosses to the south side of I-15. From Barstow to Afton Road (exit 221), it parallels but stays slightly south of I-15. Both the usually dry river and I-15 traverse the floor of Lake Manix between Yermo and Afton Canyon Road.

Riverbank exposures are not visible or easily accessible from I-15, but gullies draining to the Mojave River did erode lakebeds in two places along the highway. Eastbound travelers have the better view. One exposure is 5 miles east of Harvard Road (exit 206), where the highway dips gently into a broad swale between Mound Ditch and Flat Ditch (34°59.778N, 116°34.118W). This viewing point is approximately 150 yards south of the highway but north of the railroad tracks. Here, lakebeds are soft, fine grained, and slightly greenish. Westbound travelers should watch for this site 2.2 miles west of Field Road (exit 213) as they start down into the swale. A larger exposure of lakebeds, clearly visible to motorists going in either direction, is 3.5 miles east of the rest stop (exit 217) and 0.8 mile west of Afton Road (exit 221). Drive to this lakebed exposure via the Dunn frontage road west from the Afton exit.

VIGNETTE 15

PLEISTOCENE LAKE MANIX

Flamingos in the Desert

SAN BERNARDINO COUNTY

Approximately 3 million years ago, the Earth's climate changed from a long extended warm phase to a generally cool phase with climate oscillations that varied between cool glacial periods and warm interglacial times. The last 2.6 million years, known as the Pleistocene Epoch, featured large continental glaciers in northern latitudes and wet conditions in southern latitudes. Geologists refer to these wet or rainy intervals in geologic history as *pluvial*. In particular, the term is used most often when referring to the wet and moist glacial phases of the Pleistocene Epoch, and water bodies formed from the rain are called pluvial lakes.

The last worldwide glacial maximum, known as the Wisconsin Glacial Episode in North America, lasted from approximately 25,000 to 10,000 years ago and was accompanied by precipitation that was roughly double what we experience today. Plants and animals flourished with this extra moisture, and many rivers and streams flowed year-round, filling lake basins with bubbly blue water. Since the ice retreated about 10,000 years ago, the climate has gradually warmed and precipitation has decreased, leading to the arid to semiarid conditions we experience in the Mojave Desert today.

The Mojave Desert must have been a rather inviting place during the last glacial maximum. The land supported grass, flowers, bushes, and even small trees such as pines and junipers. The region's strong westerly winds would have made Lake Manix, the subject of this vignette, a great place for windsurfing—had any of the local denizens been so inclined.

Those pluvial lakes would have been beautiful in their prime. Their shorelines, lush with bulrushes and other plants, provided rich habitats for birds, mammals, reptiles, amphibians, fish, insects, and shelled creatures. Native Americans lived along the lakeshores. Walker and Pyramid Lakes in west-central Nevada are modern desert lakes that may give some sense of what the vanished Pleistocene pluvial lakes might have looked like.

What evidence tells us that a lake once occupied this basin? The extremely flat valley floor is suggestive, but many desert basins, never occupied by lakes, have relatively flat floors. Fortunately, the Mojave River scoured 100 to 200 feet into the floor of Lake Manix's Afton Arm, exposing ancient sedimentary beds. The uppermost layers consist mostly of thin layers of fine sand, silt, and clay—typical lake deposits. Although they accumulated in the same basin, they rest on even older lakebeds and river-deposited sand and gravel deposits that are not part of the younger Lake Manix sequence. These pre–Lake Manix beds contain at least three volcanic ash layers, the lowermost of which is the 2.1-million-year-old Huckleberry Ridge ash from Yellowstone.

A volcanic ash layer near the bottom of the Lake Manix beds correlates with a 185,000-year-old volcanic tuff (ash) from the southern Sierra Nevada, indicating that water must have flooded the basin at least 185,000 years ago. Detailed studies of wave-carved notches in shoreline bedrock show that the lake had at least eight high stands between 45,000 and 25,000 years ago. This shoreline record tells only the last chapters in a long and complicated history.

View looking southwest from Afton Road across the relatively flat floor of the Afton Arm of the Lake Manix Basin.

View looking south from Afton Road at deeply dissected, tan, older alluvium with badlands topography that lies behind the well-stratified Manix lakebeds (lower one-third of photograph), with dark-brown early Miocene volcanic rocks of the Cady Mountains on the skyline.

Hike into some of the gullies to see lakebeds adjacent to the Afton Road off-ramp as well as along the Dunn frontage road. Layers of brownish river-borne sand and gravel are interbedded with thinly bedded lake deposits of silt and clay. The river deposits are strong evidence that, at times, Lake Manix shrank dramatically and perhaps even dried up altogether.

Lake Manix shoreline features include wave-cut cliffs, terraces, ridges, and beaches left behind when the lake dried up. To see a well-preserved beach—the lake's highest and youngest shoreline—take Afton Road (exit 221) and drive south a hundred yards or so on a gravel road on top of a wide and level ridge that leads to Afton Canyon. The ridge is a magnificent beach bar built by large waves driven east down the lake by powerful westerly winds. The ridge's western edge, as well as its ridge face, is littered with souvenirs from Lake Manix: smoothly worn, flat pebbles, some nicely circular. These typical beach pebbles would make great skipping stones if we had some water.

Fossil remains of animals and birds that lived along the shores of Lake Manix are preserved within the shoreline deposits. Animals

Soft rounded mounds are remnants of lakebeds visible along Dunn frontage road about a half mile west of the Afton Road exit. (35°03.77N, 116°25.40W)

Oblique aerial view looking north at the beach ridge (at center with the road on its top) leading south from the Afton exit on Interstate 15 (upper left corner of image). Lake Manix would have been on the left (west) side of the road. Beach pebbles are well exposed in gully heads near the ridge crest.

Flat beach pebbles interlayered with sand in a small excavation near the north end of Lake Manix beach bar about 600 feet southeast of the Afton exit from Interstate 15. Pencil is 5.5 inches long. (35°04.126N, 116°24.631W)

Mostly disc-shaped beach pebbles on the ground surface once glistened beneath the water as lake waves shifted them up and down the beach. Largest pebbles are about 2 inches long. (35° 04.126N, 116° 24.631W)

represented include dogs, cats, bears, horses, camels, antelope, bison, sheep, and mammoths. Most of the noncarnivores were grazers. Shoreline birds, including storks, pelicans, cormorants, grebes, ducks, geese, eagles, cranes, and even two species of flamingos, were abundant. Imagine flamingos standing on one slim leg, solemnly surveying the Lake Manix scene today! It really happened. Under the flamingos' feet were snails, beetles, fish, and an occasional turtle.

The fossil-bearing area lies within Mojave Trails National Monument, so collecting specimens of any kind—vertebrate or invertebrate (bones or shells)—from the late Pleistocene Manix Formation is illegal except by special permit from the National Park Service.

Instead of experiencing a fine and jail term, go see splendid representative examples of these magnificent creatures in the George C. Page Museum, next to the La Brea Tar Pits in Hancock Park, 5801 Wilshire Boulevard, Los Angeles, 90036. It is well worth the visit.

Mojave River Feeds Lake Manix

By now we've probably convinced you Lake Manix existed during the last ice age, but why do you suppose it formed here? Most water for desert lakes comes from melting snow and ice in high nearby mountains, as well as from local rain and runoff. Such water fed the pluvial Mojave River 185,000 to 10,000 years ago. The San Bernardino Mountains received heavy snows that created a handful of small glaciers on San Gorgonio Mountain. Water flowing out of the San Bernardino Mountains, first north, then east, and finally north again, reached Death Valley by way of a succession of at least three large lakes connected by the Mojave River, which was a perennial stream back then. Death Valley, its floor at 282 feet below sea level, was the ultimate sink for much of our southwestern desert streams. Runoff from the Sierra Nevada ran all the way to Death Valley via a string of four large lakes connected by the Owens River. The Spring Mountains with lofty Charleston Peak (elevation 11,916 feet), 30 miles west of Las Vegas, Nevada, still feed water to Death Valley through the Amargosa River. Not surprisingly, Death Valley harbored its own pluvial lake. Lake Manly, approximately 105 miles north-northwest of Lake Manix, was more than 100 miles long and 600 feet deep (for more on Lake Manly, see *Geology Underfoot in Death Valley and Owens Valley*).

Lake Manix formed along the path of the Mojave River because various topographic impediments blocked the river's flow. The lake grew to a maximum size of 91 square miles, with three large bays, or arms. Coyote Arm, now the location of Coyote Dry Lake, was north of the present-day location of Interstate 15, tucked in behind the Calico Mountains. Troy Arm was farther south; it is currently traversed by the Santa Fe Railroad and Interstate 40. People rushing to the gaming tables at Las Vegas speed along Interstate 15 through the heart of the Lake Manix Basin and its Afton Arm.

The beach ridge along the Afton Road is part of the highest and youngest shoreline of Lake Manix. The size of the ridge indicates that the lake surface stayed at that level for a long time, presumably because the floor of the stream flowed over rock that was difficult to erode. Hard, dense rock on the floor of a stream course becomes the level below which the stream cannot quickly erode. Given enough time, enough water volume, and transported sediment, however, the stream will ultimately abrade and cut through the hard ledge.

The bedrock that held the level of Lake Manix steady was at the east end of Afton Arm, about 5 miles southeast of the Afton Road off-ramp. With time and patience, the river cut through this hard ledge. When that task was completed, water rushed over less-resistant rock and quickly carved Afton Canyon. The Mojave River was then able to discharge from Lake Manix at a lower level and flow to Soda Dry Lake (south of Baker), Silver Lake (north of Baker), and Silurian Lake (north of Silver Lake) on the way to Death Valley. All of these Pleistocene pluvial lakes are dry playas today—except for brief periods when winter storms provide enough runoff to place an inch or two of water on the normally dry surfaces.

The Mojave River flows at the surface in Afton Canyon, supporting willows and other vegetation. The railroad cuts across a meander neck here and continues down the canyon in the distance.

Ancient desert lakes were lovely oases. Those with through-flowing streams were fresh, but those with no outlets gradually turned brackish and eventually saline as evaporating water left dissolved salts behind. Late Pleistocene pluvial lakes dried up as climate changed some 10,000 years ago; in the end they all became saline and brackish, and most dried up completely. What we see today is the hot, dry, dusty arid end member of a Quaternary climate oscillation that has swung from wet to dry over the millennia.

Today, the Mojave River is a seasonal stream, but it fills much of its riverbed when the snow melts. Even so, people who live along its banks must reckon with floods when an unusually warm spring follows a wet winter with heavy snowpack in the mountains. Such floods carry water all the way to Silver Lake playa north of Baker, flooding that community en route.

Several feet of water can accumulate in Silver Lake and linger for the better part of a year. The historically recorded high water depth in Silver Lake was 10 feet during the abnormally wet winter of 1916. Water in Silver Lake must be 36 feet deep to flow over the sill at its north end and spill over into the old Mojave River channel that drains into Death Valley. There is no evidence that the Mojave River ever flowed all the way to Death Valley during recorded history, but the now-dry streambed it followed during prehistoric pluvial times, north from Silver Lake, is clearly visible in the landscape.

The Mojave River is deceptive. Coarse sand in the riverbed is so porous that water percolates into and through it easily. During dry summer months the river goes underground, where its water moves slowly downstream through its porous sand bed. This route is hard on fish, but it conserves water by slowing the velocity of runoff through the tortuous underground path. The subsurface route also reduces moisture loss from evaporation and reduces surface erosion as well.

The river comes to the surface where it flows over impermeable granitic rock at the Victorville Narrows. Within a mile or two downstream, water soaks back into the sandy streambed, except during winter floods when surface flow extends far out onto the desert floor. The narrows can be viewed at Victorville where highly fractured granitic hills rise above the surrounding terrain and CA 18 crosses the Mojave River. Interstate 15 crosses the Mojave River about 1 mile farther north.

VIGNETTE 16

THE BLACKHAWK LANDSLIDE

Eighty Seconds of Catastrophe

SAN BERNARDINO COUNTY

Blackhawk landslide, a huge elongate lobe of shattered rock, extends 4.5 miles north across the desert alluvial floor from the San Bernardino Mountains. This famous landslide is named for the mountain crest located high above it. Geologists still debate how such a thin sheet of shattered rock could travel so far across the nearly flat desert floor from its source in the mountains. They can't ask witnesses what they saw because it happened thousands of years ago.

The high abrupt face on the north side of the San Bernardino Mountains, with a broad alluvial apron at its base, is an impressive escarpment. The face is approximately 3,650 feet high, measuring from the desert floor at CA 247 (elevation about 3,100 feet) to Blackhawk Mountain with its high point at Silver Peak (elevation 6,757 feet). The sharp topographic transition from the gentle north-sloping alluvial apron to the steep mountain front is a fault-line scarp—the topographic expression of a thrust fault along which the San Bernardino Mountains were pushed up to their present height above the relatively featureless desert terrain below.

Much of Blackhawk Mountain is made up of the Furnace Limestone of probable Mississippian age (359 to 323 million years). This Paleozoic-age limestone, which has been metamorphosed in places to marble, was thrust faulted over much younger Miocene to Pliocene rock, the 7-to-4-million-year-old Old Woman Sandstone. The nonmarine sandstone was deposited on an eroded surface of really old crystalline basement—Proterozoic-age metamorphic and granitic rock. This sequence is important to remember when we analyze the landslide—Paleozoic limestone over very young sandstone over very old basement rock.

Weak sandstone beneath dense massive bedrock on the face of a steep mountain slope is a recipe for disaster because sandstone weathers and erodes rapidly, thus undermining the overlying cliff-forming marble. The geologic process, wherein groundwater seeps through and erodes porous unconsolidated sediments, is known as sapping.

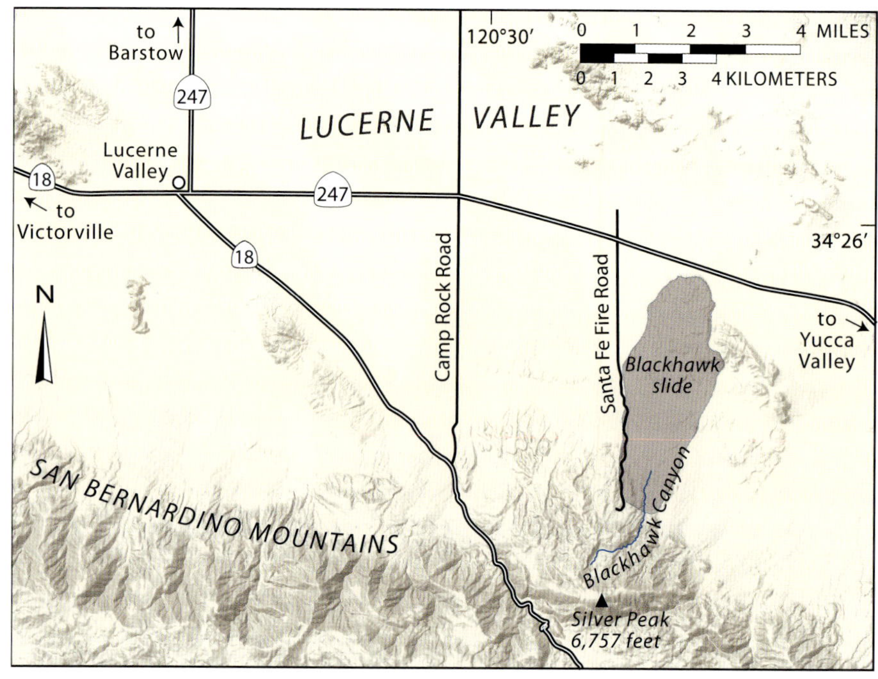

GETTING THERE

Blackhawk landslide is 31 miles east of Victorville via CA 18 and then continuing east through the settlement of Lucerne Valley on CA 247 for 10 miles. If you are traveling from Yucca Valley, take CA 247 north and west for 36 miles. From Barstow, travel south and east 42 miles on CA 247.

As a bit of a diversion, pack a picnic lunch and plenty of water and make a day-long loop by circling the San Bernardino Mountains. All roads are paved; you will see some fine high desert country and become acquainted with an unusual piece of geologic real estate. From the Los Angeles metropolitan area, you'll cross the Transverse Ranges on I-15 through Cajon Pass (Vignette 11), then at Victorville, head east through Apple and Lucerne Valleys to the landslide. You may shorten your journey by taking Bear Valley Road cutoff east from I-15 (exit 147) to a junction with CA 18 east of Apple Valley. After your visit to the landslide, proceed east and south via CA 247 to a junction with CA 62 in Yucca Valley. Follow CA 62 west and south to I-10 in Coachella Valley and then proceed west through San Gorgonio Pass between two oft snowcapped sentinels, San Jacinto Peak (10,834 feet) and San Gorgonio Mountain (11,503 feet), back to your home area.

Aerial view looking south over the Blackhawk landslide lobe, which traveled from the mountains onto and across the alluvial plain. Note transverse wrinkles near the slide's terminus. Alluvium covers the upper part of the slide. Dashed white line outlines the slide. (34°24.7N, 116°47.2W) —Photograph by Doc Searls

The relentless undercutting by sapping steepened the slope until it collapsed. An earthquake could have triggered the slide, but these more mundane erosional processes set the stage.

We know the landslide didn't occur recently because much of the upper (southern) 3 miles of the slide is mantled with alluvium washed down from the adjacent mountains. The landslide cannot be younger than about 17,000 years, according to an isotopic age obtained from freshwater snail shells collected from pond deposits in depressions on the landslide. The degree of gully dissection along the margins of the landslide, as well as the amount of alluvial debris mantling the upper two-thirds of its surface, support that age inference. It's reasonable to estimate that the Blackhawk landslide occurred 25,000 to 17,000 years ago, during the last glacial maximum when precipitation was much greater than today.

Exploring the Blackhawk Landslide

CA 247 passes about 600 feet from the north end of the slide. If you approach it from the west, don't be misled by the discontinuous mantle of geologically young windblown sand and silt that obscures the underlying slide. The slide's approximately 50-foot-high perimeter can be reached along a short gravel road that leads to two small gravel quarries at the toe of the slide. This road extends south from where a wooden-pole power line coming from the east turns north along Santa Fe Fire Road.

Vertical aerial photo of the terminal part of the Blackhawk landslide lobe.

The gravel quarries and gullies dissect the north margin of the slide and offer a good look at the slide's innards, especially of angular shattered rock fragments. Exposures are best in the walls of the west quarry. Wear a hardhat if you want to get close, and be mindful that even in these seemingly harmless quarry walls, gravity never sleeps. The walls are precarious at best and could collapse at any time. If the quarry is being worked, you may not be admitted. Proceed cautiously if it is inactive.

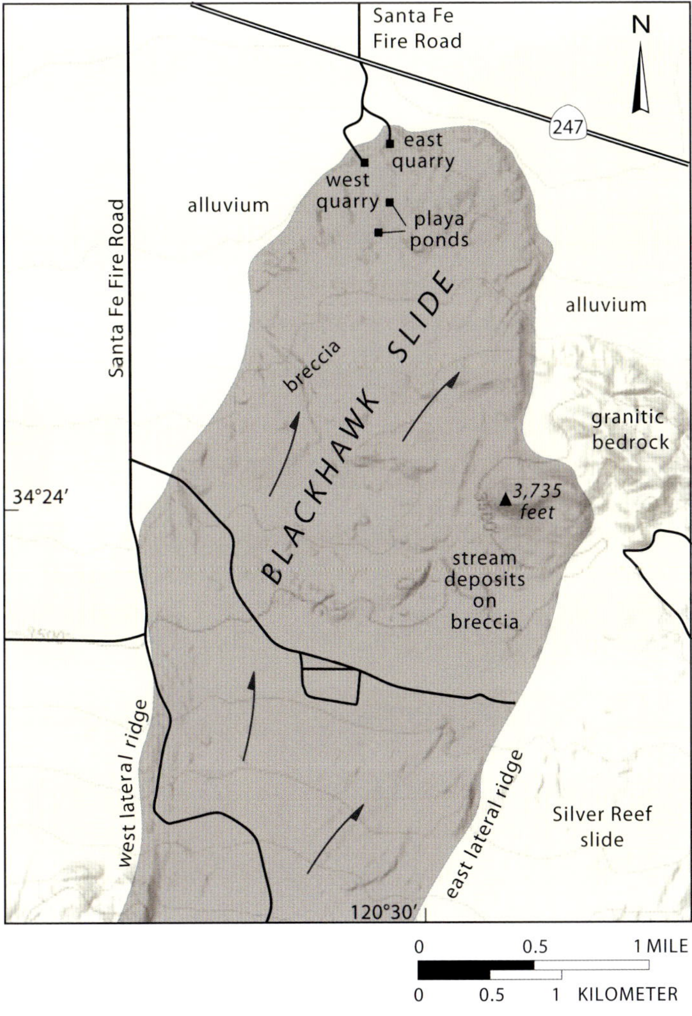

Diagrammatic rendition of the surface features on the Blackhawk slide. The landslide width varies from about 1.1 to about 1.7 miles and thickness varies from about 30 to 50 feet.

The steep walls of the quarry consist of fractured white and blue-gray marble and limestone. The source rock for the marble breccia is the Furnace Limestone. The average breccia fragment is about 1 inch across, but angular clasts up to 3 to 6 inches wide are abundant, with the prevailing size range being from powder to about 10 inches. Some larger clasts, up to 3 to 10 feet across, have been found. One giant block is 15 feet by 25 feet by 35 feet, about the size of a double-wide trailer. The landslide mass contains about 10 billion cubic feet of crushed and shattered rock. If you have ever tried to spread just

Marble crackle breccia is exposed in a steep wall of the west quarry. Staff is 4.5 feet long. (34°25.30N, 116°47.26W)

STONES AND GRAVEL, CLASTS AND BRECCIA

What most construction people and gardeners refer to as stones or gravel, respectively, have specific geologic names that refer to origin or genesis. Generically, these rock fragments are referred to as clasts. A sedimentary deposit that contains an abundance of rounded clasts is referred to as a conglomerate. Rounding implies the clasts have been water-worn and abraded by being tumbled along in a stream or river, or tossed around on a beach. Angular clasts, referred to as breccia, are not so abraded and have not traveled far, if at all.

1 cubic foot of crushed gravel (human-made breccia), you have some feel for the colossal task accomplished by the Blackhawk landslide.

The color banding and bedding in what were once large, unbroken chunks of marble can be traced many feet across small, closely nested breccia fragments, with little disturbance aside from tiny cracks coursing through what were originally solid, coherent blocks of marble. Geologists commonly describe such assemblages as crackle breccia. Marble is the predominant rock in the Blackhawk landslide mass, but breccias consisting of different rock types are also present. The presence of different rock types is an important key to our understanding of the landside's mechanics.

Scattered lenses of brownish sandstone are exposed at and near the base of the slide's terminal and lateral edges, one of which is well exposed in a roadcut leading to the east quarry. Lenses of similar rock have been found in some gully walls and within the slide itself. This sandstone breccia is derived from the Old Woman Sandstone. In a few isolated places within the landslide, small masses of brecciated Proterozoic metamorphic rocks and coarse-grained granitic rocks are exposed beneath the breccia fragments of Old Woman Sandstone.

Folds within the crackle breccia are exposed in a wall of the west quarry near the terminus of the slide. The dark-gray vertical mass is a clastic dike that formed when loose sediment intruded a fracture in the rock during the landslide event. Staff is 4.5 feet long. (34°25.28N, 116°47.23W)

This geologically ancient and geologically young suite of rocks was carried down the mountain all at once and all together in the catastrophic Blackhawk landslide without losing its stratigraphic order!

Crackle breccias, abrupt margins, surface wrinkles, and folds within the breccia exposed in the west quarry all suggest that the slide stopped abruptly. The breccias represent large blocks of rock carried as coherent pieces until the moment of impact, when they shattered into billions of tiny fragments. These fragments could not possibly have been carried as separate pieces from the source to their present site and still have maintained their original relative positions; the slide must have moved as a relatively intact mass until the very end. When it skidded to a stop, any remaining kinetic or potential energy was dissipated by cracking and shattering coherent blocks. It takes a lot of energy to form fresh cracks in a rock and to cause the shattering these rocks suffered.

Let's climb onto the landslide, make some observations, and see if we can formulate some hypotheses about how it happened. It is an easy walk from the west quarry to get to the slide's surface, or you may take any of the small gullies cut into the slide's marginal ridges. Walking on the slide's surface is easy, allowing you to roam widely. As always, watch where you step to avoid an unwelcome encounter with some innocent sleeping critter.

The top is a hummocky surface of rounded hillocks and ridges with interspersed longitudinal furrows. Scattered undrained depressions are floored with fine-grained, light-colored pond deposits that were laid down after sliding ceased. A rim-like ridge at the terminal edge rises several feet above the central slide surface; this feature is more obvious from the air than in the field. Rounded ridges extend transversely across the slide, giving its lower half a rumpled appearance, again more obvious from the air than on the ground. These transverse ridges on the slide's surface, including the terminal rim, formed by compaction and upward thrusting as the terminal part stopped while the remaining central body continued to push forward. Cessation of movement progressed backwards like a wave moving upstream.

Ridges in the upper reaches of the slide surface are fewer and tend to run parallel to the slide's path rather than across it. Two prominent ridges form lateral borders for the upper half of the slide. The approximately 50-foot-high east-side ridge is notably linear. The west marginal ridge is just as high, although more notched and ragged.

Granitic hills rise several hundred feet above the alluvial apron on the east side of the lower part of Blackhawk landslide. One hilltop, identified as Hill 3735 from its topographic elevation, was overridden

View looking southeast across a small playa deposit of white silt and sand in a closed depression on the Blackhawk landslide. Shells from such a deposit yielded an age of about 17,000 years, so the landslide occurred before then. (34°25.134N, 116°47.168W)

by the landslide, which had to have ascended at least 250 feet to make it over the top. This hill lies approximately 1.5 miles south-southeast of the two gravel quarries and is easily identified by its smoothly rounded summit and the many gullies that cut into and decorate its side slopes. It can be reached by driving approximately 0.75 mile southwest from the west quarry, and then turning east-southeast for about 1.5 miles up onto the slide surface via Zircon Road, which eventually becomes Wagner Ranch Road. At the end of the road, the hill is a short one-third mile walk to the south.

The landslide sheet was dry, both internally and at its base. There is no evidence to suggest that water discharged from landslide margins, or any suggestion of a fine-grained mud slurry at the base of or within the slide, as would be expected if the slide formed as a water-rich mudflow. The landslide probably generated a blast of air beyond its terminus, as has been reported for similar historic slides elsewhere where massive tree blow-downs occurred. Evidence to support this hypothesis is lacking at Blackhawk because any such blown-down vegetation has long since decayed away if it ever existed in the first place. Individual rocks were not ejected into the surrounding

Aerial view looking east across the hummocky surface of the Blackhawk landslide. Hill 3735, which was overtopped by the slide as it flowed from south to north, is in the middle distance.

terrain, judging from the sharpness of the landslide's edge. Except for enveloping dust, and the severe air blast, anyone standing a short distance from the terminus might have survived, although probably with newly grayed hair.

We may now draw several conclusions from our observations about the composition and structure of the landslide sheet. First, the stratigraphic sequence in the face of Blackhawk Mountain, specifically Furnace Limestone on top of Old Woman Sandstone resting on weathered and disintegrated gneiss, quartzite, and granitic rock, is preserved in the landslide, although much disturbed and thinned. Second, the bulk of the landslide moved as a sheet without significant internal mixing, as would have occurred had it flowed like a fluid debris flow. Third, the landslide climbed 250 feet over the top of Hill 3735, requiring a velocity calculated at not less than 75 miles per hour and possibly as much as 170.

We can further calculate that if the slide had a launch speed of 170 miles per hour near the head of the alluvial plain, as suggested by its fall down the face of Blackhawk Mountain, it could have accelerated to a maximum velocity of 270 miles per hour descending across the alluvial apron. The entire journey from launch to terminus would require only about 1 minute at these velocities. The run-out of 4.5 miles on a slope as gentle as 2 to 3 degrees indicates minimal drag, superb lubrication, or some exceptional mode of movement at the base of the slide.

How did this huge sheet of shattered rock move so far over a 2- to 3-degree slope? According to the most popular model, known as the air cushion hypothesis, the descending rock mass of Blackhawk landslide fell about 3,000 feet, bounced off a granite shelf, and launched several hundred feet into the air. As it settled back, the debris captured and trapped underlying air and compressed it into a basal layer perhaps only 1 foot thick. This thin layer of compressed air allowed the landslide sheet to flow across the alluvial apron. If trapping air beneath a moving slide seems unlikely, remember that the slide was moving very fast, was huge, and needed to contain the air for only a minute or so. Such a phenomenon has been witnessed during much smaller rockfalls, including the 1996 Happy Isles rock fall in Yosemite Valley (see *Geology Underfoot in Yosemite National Park*).

It is noteworthy that survivors of the massive Madison Canyon slide, which happened on the west side of Yellowstone National Park at the time of the 1959 Hebgen Lake earthquake, reportedly clung to trees to keep from being blown away. The clothes of some survivors were torn off by the powerful wind escaping from beneath the slide and by air being displaced by the fast-moving slide. These reports support the idea of the air cushion hypothesis as a mechanism for giant slides.

Fine-grained Old Woman Sandstone, weathered gneiss, and granitic rock, as well as some surface soils within the basal slide debris, may have formed a membrane that helped impede the escape of air. Heavy suspension of dust within the basal air layer may have had a similar effect—effectively becoming a dense, air-dust slurry.

The marginal ridges are high relative to the middle of the slide surface because they formed early, when the thick leading part of the debris sheet was in motion. Their height suggests that lateral margins of the slide lobe settled down while the main mass of the slide continued, thus helping to confine the air layer. Then the main mass moved on, leaving the grounded marginal ridges standing higher than the following thinner, inner part of the slide sheet.

Another hypothesis is that high-frequency sound waves, generated by the sliding mass, could keep the particles of a fine-grained basal layer separated, thus creating a fluid-like condition that permitted easy and rapid movement. The slide sheet simply went along for the ride on the back of a thin, fluidized, basal layer of fine particles.

This fluidization concept merits consideration, although it does not explain some features of the Blackhawk slide as neatly as the air cushion mechanism. The acoustical sound wave cushion model works better for large, presumably high-velocity slides on the Moon's surface, where there is no atmosphere, and it may also apply to the huge slides of Mars, where the atmosphere is less than 1 percent as dense as that on Earth.

Pomona College geologist A. O. Woodford and his associate T. F. Harriss first published their discovery of the Blackhawk landslide breccia in 1928. Many geologists have subsequently poked around, admired, and commented on it, but the most comprehensive study and the suggestion that it may have flowed on a cushion of compressed air came from Ronald L. Shreve, a former UCLA professor of geology and geophysics. His research paved the way for H. Jay Melosh's alternative suggestion that acoustical fluidization might have provided sonic cushion for Blackhawk as well as for similar slides on the Moon and Mars.

Could it happen again? Remnants of past landslides exist along an approximately 2-mile-wide and 3- to 4-mile-long, landslide-scarred range front east of the Blackhawk slide, indicating that this area is prone to landslides, and thus another can certainly happen again in the future. The basic conditions favorable to sliding and the processes creating over-steepened slopes are still at work, but the wetter climatic conditions when Blackhawk failed are, for the most part, absent today. So, it is not likely that a repeat catastrophe will occur while you are out there exploring this extraordinary example of geologic processes at work. You are probably more likely to be surprised by a nearby meteorite strike than to witness the next edition of a Blackhawk-like event.

VIGNETTE 17

KELSO DUNES

A Huge Pile of Living Sand

SAN BERNARDINO COUNTY

Exploring sand dunes is fun. They can be mysterious and beautiful. Many seem alive, changing shape with each shift of the wind. The witching time for dunes is just before sunset and just after sunrise, when shadows are long and deep, and when one may appreciate the remarkable grace of curving dune crests and the complicated intermingling of geometrical forms. Walking through dunes at sunset, when the western sky reflects its twilight glow onto the sand, can be a mystical, magical, almost religious experience. Best seasons to visit the dunes are winter, early spring, and late fall when it is not too hot.

Kelso Dunes aren't the highest or most extensive in the California desert, but they are among the most spectacular. They have been closed to off-road vehicles since 1973, so they are relatively protected from the kind of damage that has reduced some of California's other dunes to barren piles of sand.

The dunes rise from a broad alluvial apron that slopes gently down to the north from the Granite Mountains. Projection of this smooth surface under the highest dunes suggests a sand thickness of at least 700 feet. This 45-square-mile patch of marvelously sculpted sand dunes lies at the end of an umbilical cord of windblown sand 4 to 5 miles wide that extends 35 miles east and southeast from Afton Canyon (see Vignette 15) and across Devils Playground, 10 miles northwest of the dunes, which itself receives additional sand from the north. The rugged Providence Mountains to the east, which supply little sand, were named by early Mormon travelers in appreciation for springs found there.

Prevailing winds in the Mojave Desert blow from the west. Here at Kelso Dunes, the shape of the landscape channels the winds into the prevailing west-northwesterlies, although ripple marks and dune forms indicate that strong winds may attack the dunes from any direction. These storm winds cause rapid and profound changes in dune shapes.

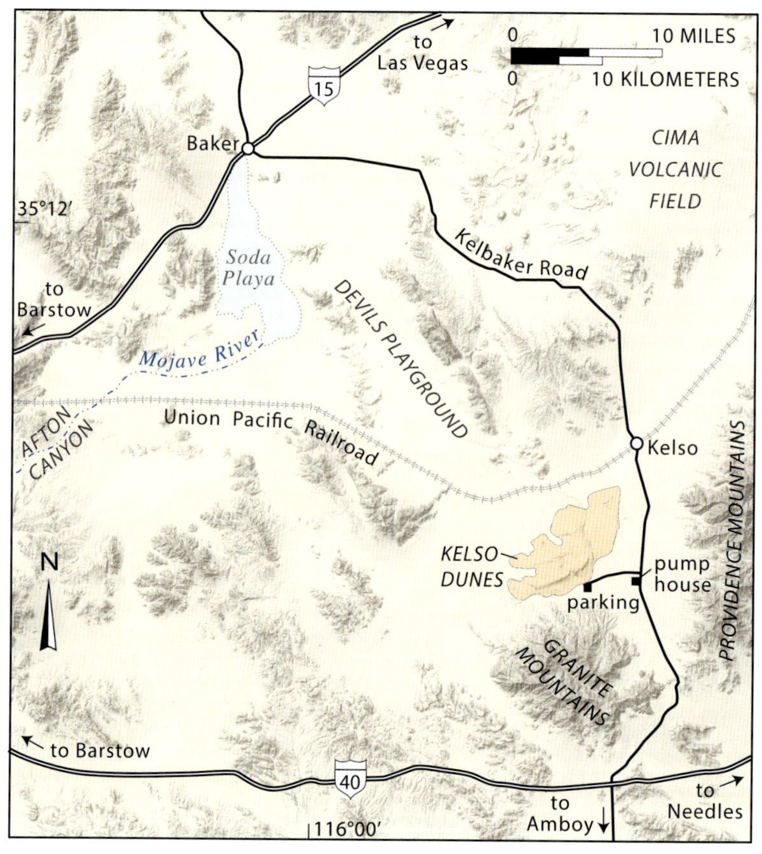

GETTING THERE

Kelso Dunes are located in the eastern Mojave Desert about midway between the diverging paths of Interstate 15 to the north and Interstate 40 to the south. To get there from Barstow, take I-40 east about 75 miles to its intersection with Kelbaker Road (exit 78), thence north about 12 miles from the I-40 junction. Access via I-15 is 44 miles longer, proceeding southward from Baker via Kelbaker Road (exit 246 from I-15). That route skirts the geologically young Cima volcanic field with its 40 cinder cones and lava flows and passes Kelso (35°00.741N, 115°39.229W), 7.5 miles northeast of the turnoff to the dunes. At Kelso is a restored Union Pacific Railroad depot of unusual elegance, size, and history. It was established in 1906 along the San Pedro, Los Angeles, & Salt Lake Railroad, which later became a subsidiary of the Union Pacific. Potable water, restrooms, a small museum, and a bookstore are at Kelso Depot Visitor Center.

View of Kelso Dunes looking north. The highest point on the left skyline is 500 feet above the photographer. (34°54.2N, 115°43.1W)

Four large, linear ridges that bear east to northeast are the main features of Kelso Dunes. Many smaller transverse ridges cover their flanks and the areas between them. A typical transverse ridge is 200 to 300 feet long. Its broad windward flank slopes 10 to 15 degrees and has a firm surface commonly decorated with little ripples. The windward slope climbs to a smooth crest, which drops off into a steep leeward face that stands at the angle of repose of dry sand, up to 34 degrees. The upper edge of the steep lee face generally ends a bit lower than the dune crest, which typically is gently rounded. Lee face heights range from a few feet to 30 feet and vary along a dune's length. The somewhat sinuous crest line alternates between broad, rounded summits and open intervening saddles.

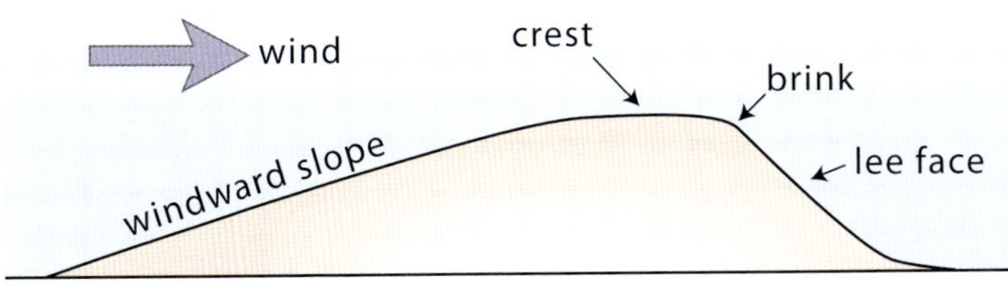

Cross section of a typical transverse dune ridge.

Wind rolls some sand grains along and bounces others across the dune surface in a series of short hops. The bouncing grains knock others across the surface as they crash back onto the dune. Those impacts kick large grains along in a series of sudden jerks, like a child kicking a can down the sidewalk. Bouncing, the most efficient means of wind movement of sand, accounts for about three-quarters of the total sand transport. This bouncing mode of sand movement is known as *saltation*, a word from Latin meaning "to leap" or "to dance" (and used to name the Italian delicacy, *saltimbocca*, which means "jumps in the mouth"). The difference between rolling and bouncing resembles the difference in mobility between a child that crawls and one that walks. The latter covers more ground and gets into twice as much mischief.

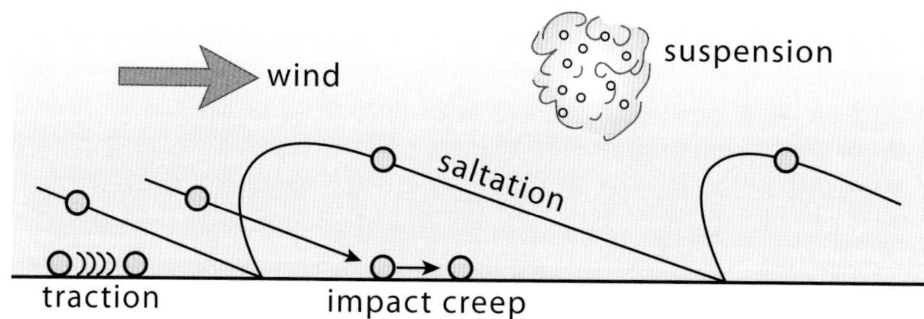

Modes of wind-driven sand transport.

Sand grains blowing up the windward slope of a dune enjoy a free ride at the expense of the wind. Bouncing sand grains cover distances up to several feet with each hop, angling back into the dune at nearly the speed of the wind. When a strong wind blows across a dune, the cloud of bouncing sand grains blurs the view of its crest and windward slope. The cloud of sand is rarely more than 12 to 16 inches high, so you can stand comfortably in it, and someone with bare legs can tell you how high the grains are bouncing.

Sand moving up the windward slope surface reaches a brink where the lee face drops off. There the bouncing grains make their last leap and are unceremoniously dumped down the lee face, coming to rest somewhere below, sheltered from the wind. Avoid the lee side of the dune, where a deluge of falling sand grains will rain upon you. The net effect of removing sand from the windward slope and depositing it on the lee slope causes the dune to migrate downwind.

The cloud of bouncing grains is denser near the dune surface, so you can be sure that most grains are bouncing to low heights and consequently making short leaps. As a result, most of the bouncing grains that cross the crest land on the upper part of the lee face, making it steep. When the slope reaches an angle of about 34 degrees, the upper part of the lee face slides in a tongue of avalanching sand toward the bottom of the slope. Those avalanches occur repeatedly on the lee face of a rapidly advancing dune when strong winds are moving sand. If you visit dunes shortly after a storm wind, you will find it easy to generate avalanches by stomping along the upper part of the lee face. It is fascinating to watch them ooze down a lee slope like syrup. Avalanches move nearly all sand on lee slopes.

A transverse dune, with its crest at right angles to the wind, is an efficient sand trap; very little sand escapes it. Wind that has passed over a dune, having lost much of its sand, starts to pick up sand from the succeeding hollow and from the lower windward slope of the next downwind dune. By the time it gets part way up the windward slope of the succeeding dune, the wind is loaded with sand and ready to deposit part of its burden. This type of deposit is called accretion sand. Experienced dune hikers know that accretion sand is made firm

The lee face of a transverse dune after a heavy wind, with a fresh avalanche tongue triggered by a footstep.

by the impact of bounding grains. It is hard to leave a footprint in it. The upper part of the windward slope has the thickest and firmest mantle of accretion sand. Near the brink of a rapidly advancing dune, however, accretion sand may be so thick that a hiker will break through into the soft avalanche sand on the lee slope beneath. Avalanche sand is so loosely packed that the hiker may sink in up to the ankles. It is smart to keep on, or a little windward of, the dune crest for the easiest walking.

Like the sand grains in dunes everywhere, those in the Kelso Dunes are smoothly worn and nicely rounded. Look at the grains under a strong hand lens or microscope and you will see that the almost perfectly round surfaces are covered with tiny pits caused by the impact of one grain against another. Grab a handful anywhere, and you see that the grains are all about the same size, a result of the powerful winnowing force of the wind.

Most of the sand consists of the minerals quartz and feldspar. Thin layers of black sand, mostly the magnetic iron oxide mineral

Tracks on rippled sand surface.

magnetite, catch your eye, especially where they make a dark patch on the surface rather than just a streak. Children enjoy dragging a magnet through the black sand, covering it with magnetite hair. Kelso sand is so rich in magnetite that a Texas entrepreneur once staked parts of the dunes as placer claims and planned to separate the magnetite to sell to a steel company. Fortunately for the dunes, this project languished.

Wind ripples are among the most intriguing features on dune surfaces. They are miniature models of transverse dunes, a few inches from crest to crest and a fraction of an inch high, and like dunes, they are asymmetrical, with a long, gentle windward side and a short, steep lee face. They make interesting patterns around surface irregularities and obstacles, such as bushes. These patterns show that the shape of the surface strongly affects wind currents along the ground, causing the currents to depart considerably from the prevailing direction. Most times, active ripple crests on dune surfaces are perpendicular to the general direction of the wind.

Ripples form when the wind is strong enough, about 25 miles per hour, to raise a cloud of bouncing sand grains. Impacts of incoming bouncing grains drive the larger creeping grains up the windward slope of the ripple. Some linger in the ripple crest; others tumble down the steep lee face.

A wind blowing 30 miles per hour can drive sand ripples along at a speed of several inches per minute. To demonstrate this, take some toothpicks or medium-sized nails into dunes on a windy day, stick them into several successive ripple crests, and watch the ripples migrate. Another fun experiment is to erase the ripples in a patch of sand 2 or 3 feet square by smoothing with your hands and watch the wind make new ones in a matter of minutes.

Dunes provide an unusual ecological niche in the harsh desert environment because they save water. Their porous sand absorbs almost all the rain that falls on them. After a rain, evaporation from the upper few inches of sand produces a dry layer that insulates the moist sand below from heat and evaporation. Months after any rain, a hole dug several feet into soil adjoining dunes will be bone dry, whereas a hole dug on the windward side of a transverse dune will usually penetrate moist sand within a few inches, at most within a foot or two. At any time of year, strong winds may blow dry sand off the dune surface exposing moist sand beneath.

A few dozen clusters of desert willow trees, well up on the south face of Kelso Dunes not far east of their highest point, demonstrate the availability of water. Some trees are dead, but many more are living, blossoming, and making seedpods—all signs of good health. Individual tree trunks approach 1 foot in diameter, and the height can reach 20 feet. The willows grow right out of dune sand, which is hundreds of feet thick. Trees probably colonized the dunes under moister climatic conditions than the present, but their offspring survive because the dunes hold water.

Burrowing animals, reptiles, and bugs know they have a temperature-controlled system in the dune sand. By burrowing to a chosen depth, they select a comfortable temperature and humidity. One of the most interesting denizens of the dunes is a little lizard that loves to lie burrowed in sand. The pressure of a passing hiker's foot inspires it to boil out of its burrow and take off across the surface like a streak of lightning. It can disappear before your eyes, either by stopping and holding still—its coloration providing almost perfect camouflage—or, more likely, by burrowing into the sand. The lizard knows the difference between accretion and avalanche sand. When wanting to escape in a hurry, it heads for the lee face of a dune and literally dives into the loose avalanche sand. Children love these astute creatures.

Inspection of dune surfaces in early morning after a windless night reveals a world of tracks and trails etched on the soft sand. A lizard leaves tail streaks between its footprints. A sidewinder rattlesnake leaves a distinctive series of cuspate curves. You have to see a beetle in action to appreciate how its track is made, and that is easy because they are active in the daytime. Tracks of larger animals, such as rabbits, foxes, and coyotes, also show up well in the sand.

Native Americans camped near the Kelso Dunes in bygone days. Artifacts such as arrow points are still found occasionally, and large grinding stones were once abundant in the vicinity. Fire-reddened stones mark sites where Native Americans had campfires. They reputedly washed blankets and furs by impregnating them with sand and then shaking them out to remove grease, dirt, and vermin.

Native Americans carried the rocks for their fire rings into the dunes, which are normally quiet and peaceful places to be. Occasionally, however, the dunes break the silence with a deep, low-pitched, booming sound like the moaning of a faraway diesel locomotive or of a small airplane. The phenomenon, known as "singing" or "booming" dunes, has been recognized in the Middle East for more than 1,500 years and in Chinese literature from as early as the ninth century. Both Marco Polo and Charles Darwin described singing dunes.

Only about thirty dunes in the world are known to boom; seven of them are in the United States, including Kelso Dunes. Scientists attribute the sounds to the internal shearing of sand grains avalanching down lee slopes. You can make the noise by vigorous kicking at a sharp dune crest to dislodge a surface layer of sand down a lee face after a strong wind, thus starting a large sand avalanche. The sand must be very dry and the air must be low humidity. Well-rounded, frosted sand grains derived from long-distance transport, or from multiple generations of sand movement because of variable wind direction, appear best-suited to the production of booming sounds.

So, where did all the sand at Kelso Dunes come from? About 35 miles west, the Mojave River flows east out of narrow Afton Canyon to build a broad alluvial plain into the Devils Playground. Every time it floods, the river renews the supply of raw, loose rock debris, much of it sand, onto the surface of this plain. The prevailing westerly winds blow the sand off the alluvial plain, east to Kelso Dunes. The uncommonly large amount of magnetite in Kelso Dunes probably comes from Afton Canyon, where mines have produced the mineral in commercial quantities. Strong winds blowing out of Soda Dry Lake near Baker probably contribute additional sand.

Why are the dunes so far out in Kelso Valley, rather than tucked up against one of the bordering mountain ranges? The location of Kelso Dunes in a valley is by no means unique. The huge dunes in Eureka Valley and Death Valley also lie in valleys (see *Geology Underfoot in Death Valley and Owens Valley*). A long study of the behavior of transverse dunes in the Kelso Valley provided a possible explanation. It found that individual transverse dune ridges moved back and forth cumulative distances well in excess of several hundred feet in ten to twelve years, but ended up within a few feet of their initial position. During this same period, the wind removed and redeposited several hundred feet of sand on dune crests, with an almost perfect balance between accumulation and erosion. Although the dunes have been very active, they have not moved far in any direction.

The ability of wind to transport sand increases with approximately the cube of its speed, so doubling the wind speed increases its carrying power by a factor of eight. Occasional very strong storm winds from the south, north, or east are able to balance the effects of the more prevalent but usually gentler winds blowing from the west.

What caused the dunes to start to grow about where they are now? One possibility is that some perturbation on the valley floor, such as a low, rough patch of bedrock, localized the initial accumulation of sand. A seemingly more likely explanation, in view of the fact that other large dune masses also prefer valley floors, is that their location reflects a balance within the complex of conflicting wind patterns. The surrounding mountain terrain could play a part in creating such a node.

View looking east of Kelso Dune ridge. The Providence Mountains form the skyline.

When we ask how old the dunes are, we mean when did great piles of sand start to accumulate at this locality. Unfortunately, datable material has not been found within the dunes; they contain no internal evidence of their age. If we knew how much sand is now being added to the dunes every year and divided that into their total volume, the result might be considered a very crude estimate of their age. But it would not be particularly reliable because the rate of sand accumulation has almost certainly varied with climatic changes over the millennia. A better approach is to ask how long the Mojave River has been spreading fresh debris over the alluvial plain at the mouth of Afton Canyon. The river gets its water from the high San Bernardino Mountains. During the last glacial episode, those mountains certainly shed much more water than they do now, so the Mojave River must have been much larger then. We know it supplied most of the water to maintain a large lake in the Manix Basin upstream from Afton Canyon (Vignette 15). This lake is thought to have started overflowing through Afton Canyon channel roughly 14,000 years ago. That could have begun the creation of the alluvial plain and subsequently of Kelso Dunes.

VIGNETTE 18

MITCHELL CAVERNS

A California Rarity

SAN BERNARDINO COUNTY

Limestone caverns are curious, spooky, and awe-inspiring features. It seems a shame that California has so few of them. The necessary ingredients for cavern formation are fractured carbonate rocks and a copious supply of water. Carbonate rocks (limestone, dolostone, and marble) are scarce in California, except in eastern California. There you will find Mitchell Caverns in the Providence Mountains State Recreation Area within the Mojave National Preserve.

When visiting Providence Mountains State Recreation Area, be sure to visit the small museum and walk the Mary Beal Nature Study Trail, a half-mile tour of the flora and fauna of the area. The museum walls are made from an astonishing array of rocks, all of which Jack Mitchell, the caverns' early developer, claims to have collected from the surrounding mountains. One can spend a pleasant afternoon just admiring these rocks. The museum steps, at an elevation of 4,300 feet, make a great place to gaze out across the eastern Mojave Desert. Descriptive plaques identify many of the nearby mountain ranges—and some that are not so nearby. On a clear day, you can see mountains in Arizona, including the Hualapai Mountains, 100 miles away.

A colossal eruption of ash-flow tuff from a caldera in the Woods Mountains, about 12 miles northeast of the caverns, covered much of the region about 16 million years ago. The tuff lapped onto the Providence Mountains and caps Wild Horse Mesa, 7 miles to the north-northeast. The tuff is well exposed at Hole-in-the-Wall Campground on Black Canyon Road, about 10 miles north of the intersection with Essex Road. Careful study of the flanks of Wild Horse Mesa reveals profiles of hills and deep valleys buried beneath the tuff.

A ranger will guide you along the broad, 0.5-mile-long trail south from park headquarters to the caverns. The first outcrops you encounter are limestone that is part of the Bird Spring Formation of Pennsylvanian and Permian age, roughly 300 to 250 million years old. Although the formation is more than 2,000 feet thick in this area, the caverns are only in the lower few hundred feet.

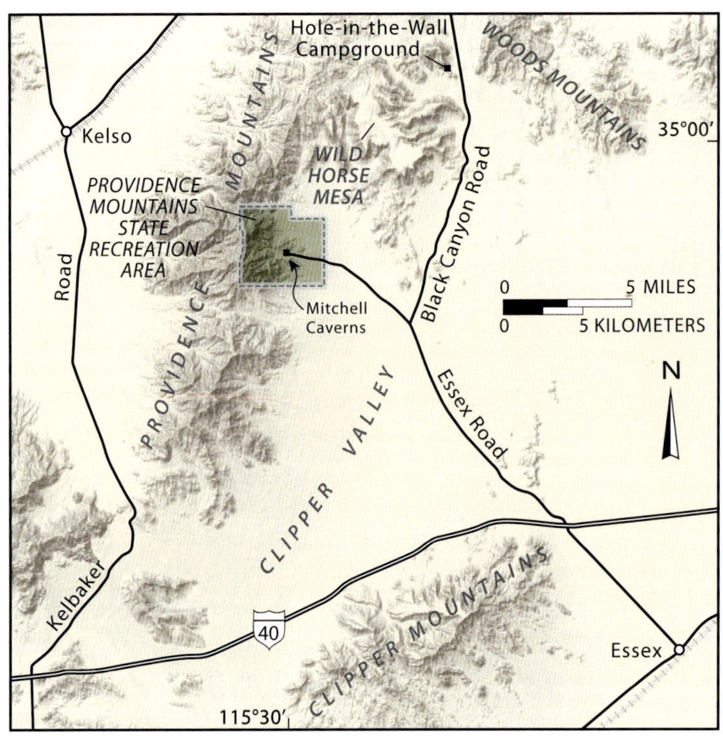

GETTING THERE

Mitchell Caverns are in the eastern Mojave Desert. From Barstow, drive 102 miles east on I-40 to Essex Road (exit 100). Follow Essex Road 9.7 miles north to its junction with Black Canyon Road. Bear left at this Y intersection and continue another 5.9 miles to the park headquarters (34°56.63N, 115°30.81W).

Mitchell Caverns, a National Natural Landmark, is the central attraction of the Providence Mountains State Recreation Area, a 5,900-acre reserve on the east slope of the Providence Mountains. Park personnel conduct fee-based, twice-daily tours of the electrically lighted El Pakiva and Tecopa Caverns by reservation only at 11 a.m. and 2 p.m. from mid-September through mid-June. The nearest food, lodging, and gasoline are 56 miles away. The nearest camping is about 15 miles away at Hole-in-the-Wall Campground. The caves remain at about 65°F year-round, so if the day is warm, you may want to take a sweater for the tour.

View from the park headquarters looking northeast to the flat-lying, interlayered tan tuff and rhyolite lava flows of Wild Horse Mesa. The craggy lower slopes of the mesa are older hills that were buried by the tuff of Wild Horse Mesa.

The gray limestone contains bits of orange, tan, and chocolate-brown chert, a finely crystalline variety of quartz. The chert exists in scattered, irregular blobs and in veins lining fractures. This association of chert blobs with limestone is common for rocks of this age; the blobs formed from silica that accumulated with the limy sediments on an ancient seafloor. The chert stands out in relief on the surface because limestone is slightly soluble in rainwater, whereas chert is not. Jagged pieces of chert on the outcrop quickly wear out the soles of your boots and easily tear your clothes (and skin, if you're not careful).

After another few steps, beyond the chert outcrops, you'll see a distinct change in the color and texture of the rocks above the trail. For several yards, the rock is pale brown and smooth, and contains large circular bands. This is a patch of flowstone, a sheet-like deposit of calcite in the form of travertine, which forms where water flows down the walls or along the floors of a cave. Flowstone lines many caverns; this exposure may mark the top of a cave room that was exposed by trail excavation. Is it possible that a large open chamber exists directly under your feet?

Look for fossils in the limestone along the trail, but keep in mind that fossil collecting is prohibited because this is a state reserve. The

Tan and orange chert blobs in gray limestone of the Bird Spring Formation. Pencil is 6 inches long.

THE GREAT DYING

The fossils you're walking over were part of an Earth that changed radically at the close of the Permian Period, about 252 million years ago, when the worst mass extinction in Earth history killed off most life on the planet, including over 95 percent of marine species. Fusulinids didn't survive the event, known as the Great Dying, and only a few species of brachiopods and crinoids limped through. The mass extinction was caused by rapid eruption of unimaginable amounts of basaltic lava in Siberia—enough to cover the Lower Forty-Eight States to a depth of 500 feet and to shroud the entire planet in sunlight-blocking gas and clouds for years!

One-celled animals called fusulinids, typically about 0.5 inch long, were common in the sea during deposition of the Bird Spring Formation about 300 to 250 million years ago. Look for their remains in limestone along the trail. —Drawing by Mary Olney

Bird Spring Formation contains the remains of diverse animals that lived in seawater during Pennsylvanian and Permian time. Brachiopods resemble clams in having paired shells, but they are not actually related to them. Dark-gray cross sections of their shells are exposed in the rock surface. Other fossils include fusulinids, the remains of floating, one-celled animals that are about the size and shape of a pine nut. Crinoids, sometimes called sea lilies, are really animals, even though they anchor themselves to the seafloor by long stems made of calcite. Loose stem pieces about 0.25 inch in diameter look like miniature coins with holes through them, or petrified washers.

Pits and grooves etched in the limestone near the first diversion culvert show the dissolving effects of slightly acidic rainwater. A small entrance to a mining adit, not open to the public, is at the second diversion culvert. The public entrance to the caverns is a little farther

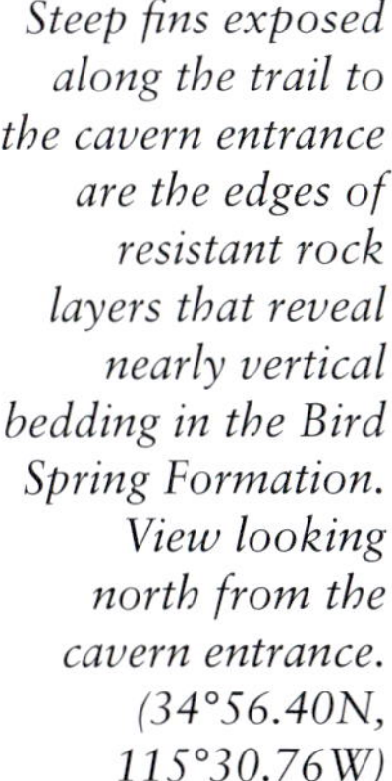

Steep fins exposed along the trail to the cavern entrance are the edges of resistant rock layers that reveal nearly vertical bedding in the Bird Spring Formation. View looking north from the cavern entrance. (34°56.40N, 115°30.76W)

along the trail, around a right-hand bend and over a bridge across a gully. When looking at the cave entrance from across the gully, you can see that the sedimentary bedding, indicated by steep fins on the rock face, dips to the east at an angle of about 80 degrees. At a point across the gully from the public cavern entrance, bedding in trailside outcrops is evident by the alignment of thin chert lenses in the rock.

About 5 yards past the end of the wooden bridge, look for nicely preserved fusulinids, crinoid stems, and brachiopods in the dark-gray limestone. Fusulinids abound in rocks near the cavern entrance.

The Formation of Mitchell Caverns

The cavern entrance is a good place to think about how caves develop. We have found the required limestone; now what role does water play in cavern formation? Rain reacts with atmospheric carbon dioxide to make carbonic acid, which makes rainwater slightly acidic. The club soda you buy in the grocery store is a more concentrated solution of carbonic acid. Polluted air normally contains nitrogen and sulfur oxides, which also react with rain to make acids. That is why rain downwind from industrial or heavily populated areas is much more acidic than normal rain—hence the term *acid rain.*

Limestone contains carbonate, a carbon atom bonded with three atoms of oxygen, or CO_3. It reacts with acids to form carbon dioxide gas and water. Squirt a bit of hydrochloric acid on limestone and watch the rock fizz as its calcium goes into solution and carbon dioxide gas escapes. The same thing happens in a more subdued way when rain falls on the rock.

Limestone dissolves readily in the abundant rain of wet regions, and valleys form. In dry regions, however, limestone resists erosion and normally stands topographically high as ridges and cliffs. But even there, exposed limestone surfaces dissolve in the occasional rain, becoming rough, pitted, and microscopically sharp.

The high peaks of the Providence Mountains are reddish-brown igneous rocks, unlike the gray limestone at your feet. These rocks intruded and deformed the Bird Spring Formation during Jurassic time, about 160 million years ago. Rocks downhill from the trail are also igneous, but they were faulted against the limestone instead of intruding it. Thus, the caverns are in a thin wedge of limestone that was faulted against igneous rocks on the east side and intruded by them on the west. Those events left the limestone badly fractured and vulnerable to the dissolving effects of acidic rainwater. The nearly vertical bedding surfaces and the many fractures allow water to move

The trail enters the right (west) entrance to El Pakiva Cavern. The left (east) entrance is sealed. (34°56.36N, 115°30.75W)

easily into and through the rock, exposing it to dissolution. Boxy networks of chert highlight some of the fractures.

Caves form where groundwater, channeled through fractures in limestone, dissolves large holes in the rock. As a fracture enlarges to a critical width of about one-quarter of an inch, water flow becomes turbulent, greatly enhancing the solution of the rock. Such large fractures thus enlarge even faster, robbing adjacent fractures of water. A few fractures eventually channel most of the water and develop into caverns.

Mitchell Caverns are several thousand feet above the floor of Clipper Valley, so the water table is well below the caverns. Why are they here? They probably began to form millions of years ago when the volcanic rocks on Wild Horse Mesa lapped against the Providence Mountains. At that time, the ground surface was above the level of the caverns, and the water table could have been above the caves. Erosion has since exposed the mountainside. Faulting has also lifted the caverns above the valley floor. Or perhaps the caverns formed after the mountains rose but when rainfall was abundant enough to maintain a throughgoing flow of groundwater, presumably during the ice ages. The last

FORMATION OF MITCHELL CAVERNS

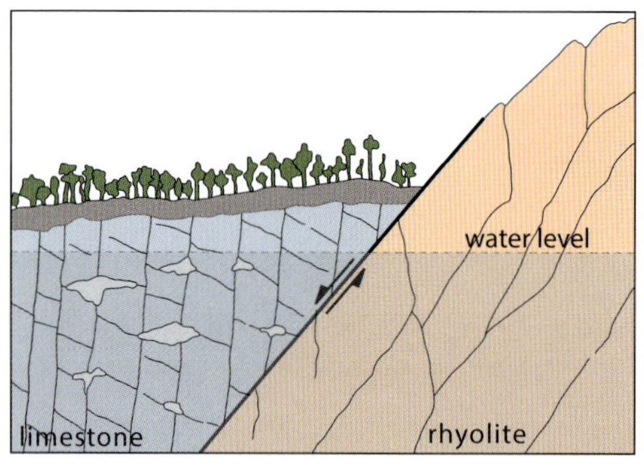

Phase 1. After the limestone of the Bird Spring Formation was uplifted, tilted nearly vertically, and faulted against the 16-million-year-old rhyolite, water penetrated along fractures and began the process of cave formation.

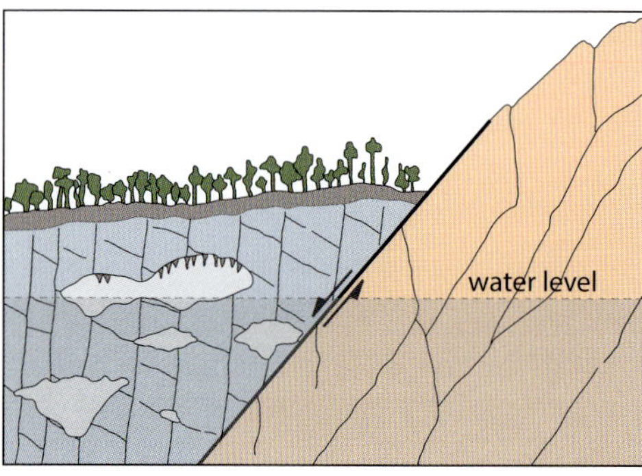

Phase 2. A luxuriant rain forest covered the area about 12 million years ago. The rainwater, rich in carbon dioxide from decomposing vegetation matter, expanded the fractures and pockets in the limestone. Eventually, water filled the subterranean chambers, and passageways formed.

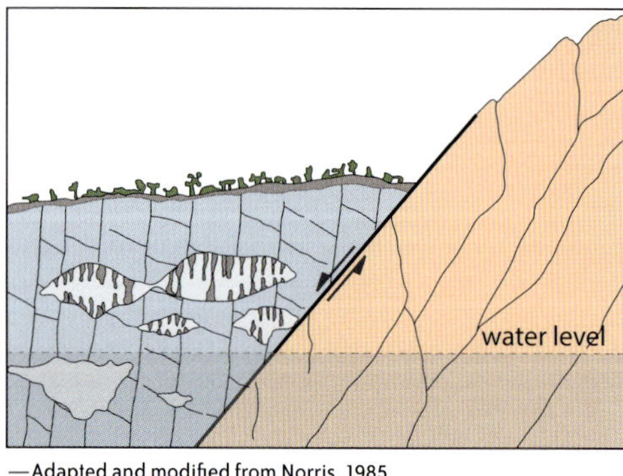

Phase 3. Climatic conditions in the region changed and the region transformed into a desert. Water level dropped below the cave level, and many thousands of years were required to sculpt and ornament the beautiful underground stalactites and stalagmites.

—Adapted and modified from Norris, 1985

great ice age, which ended about 10,000 years ago, was probably the last period when rainfall was heavy enough to enlarge the cave.

Speleologists, who make it their business to study caves, think that most cavern enlargement occurs when a cavern lies a few feet below the water table. They point out that many caverns are nearly horizontal, as are water tables, even if bedding in the enclosing rocks is inclined, as it is at Mitchell Caverns. Also, the zone just below the water table is a favorable site for dissolution because when downward-percolating rainwater mixes with groundwater, the resulting mixture is undersaturated in calcite. If the water table drops, caverns begin to fill with the ornate dripstone deposits that make them so bizarre or attractive, depending on your point of view.

Dripstone deposits form where groundwater seeps from the ceiling or walls of a cavern. Groundwater dripping into a cave loses some of its dissolved carbon dioxide to the cave's atmosphere, becoming less acidic and precipitating part of its load of dissolved calcite, thus forming dripstone deposits.

The most familiar dripstone formations are the stalactites that hang like stone icicles from the ceilings of many caves. They form where water droplets from the ceiling lose carbon dioxide to the atmosphere and deposit a film of calcite around the surface of each drop. When the falling drop breaks through that film, it leaves a thin ring of calcite attached to the cave roof. Successive drops add more minute rings of calcite, one after the other, eventually building a thin stalactite that is hollow, like a soda straw. Typically, it takes something like twenty years of fairly steady dripping to add 1 inch to the end of a stalactite. If water flowing across the ceiling encounters a soda straw stalactite,

A dense group of rapier-like stalactites emerges from cracks in the cavern ceiling. The longest is about 4 feet.

it will flow down the outside of the tube, enlarging it by depositing calcite on its outer surface.

Water that drips from the tip of a stalactite and splashes onto the floor will build a mound of calcite—a stalagmite, which is a solid column rather than a hollow tube. If the stalactite and stalagmite meet, they form a column from floor to ceiling.

Mitchell Caverns is rich in stalactites and stalagmites and contains a diverse collection of other deposits, including sheets of flowstone deposited by water flowing down the cave walls. Helictites, irregular and branching stalactites made of needle-form calcite and aragonite, are the most delicate and fragile of cave formations. Their origin is obscure and a matter of continuing debate among speleologists. Be sure the guide points them out to you.

The most prized formation in Mitchell Caverns is a large travertine shield, a semicircular sheet of calcite a few feet across and an inch or

Floor-to-ceiling limestone columns in Mitchell Caverns.

The travertine cave shield is about 3 feet across.

two thick. Shields form when water seeps slowly from a crack in the cavern wall, building layers of calcite outward.

All of the formations in Mitchell Caverns formed long ago when the climate was wetter. These days, periods of exceptional rainfall merely dampen the cavern walls but probably do little to enlarge the cave or to add to its dripstone decorations. The cave walls are dry most of the year.

The Cave Tour

The following notes and maps are meant to supplement the guided tour provided for a fee by the park personnel. The tour actually goes through two separate caves, El Pakiva and Tecopa, which were joined by a tunnel drilled in 1969–70. Depending on the source, *el pakiva* translates to "pool of water from the eye of the mountain" or "devil's house." *Tecopa* was the name of a Shoshone chieftain. El Pakiva is about 200 feet long; the main chamber of Tecopa is about 150 feet long.

The tour starts at the west entrance of El Pakiva because the adjacent eastern entrance has been walled off. After climbing steps and passing through the gate, the path curves left into a large room marked by large chunks of rock, called breakdown blocks, that have fallen from the walls or ceiling. A sharp right turn brings you into the

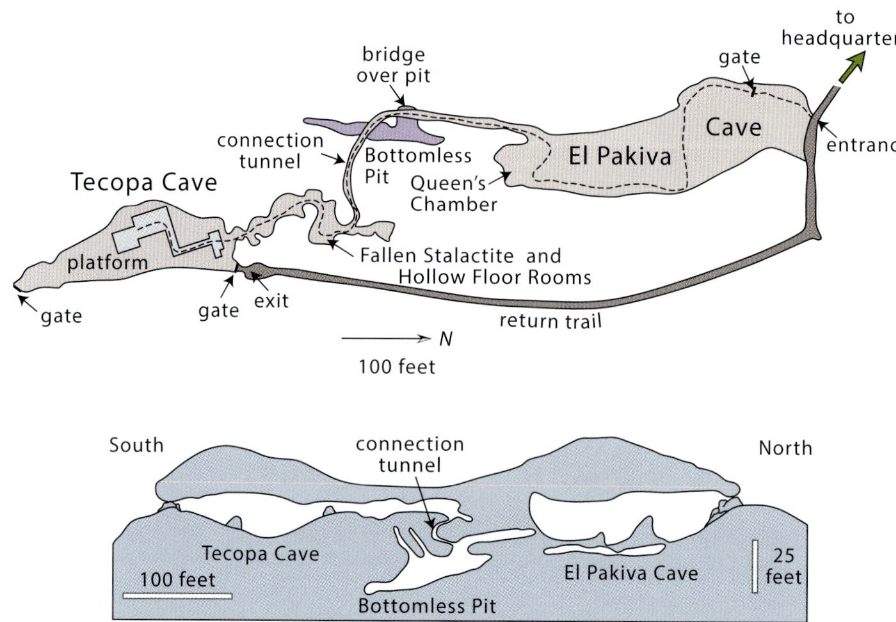

Map and diagrammatic cross section of El Pakiva and Tecopa Caverns.
—Modified from Aalbu, 1989

main chamber of El Pakiva. Its ceiling is 40 to 50 feet above you, and the trail hugs the cave's inclined east wall, which is a limestone slab that leans out over the flat floor. The opposite (west) wall, on your right, is mostly flowstone.

A walk of about 100 feet brings you to the Queen's Chamber, a magnificent room decorated with thousands of stalactites, stalagmites, and columns. There is much to see here, and your guide will point out the more obvious wonders. The path then bends to the right and climbs out of the Queen's Chamber into a narrow natural passageway that leads to a lower level of El Pakiva. In another 100 feet, the trail comes to a bridge over a lower level, which is known as the Bottomless Pit. Shortly thereafter, the trail enters the artificial tunnel that connects El Pakiva and Tecopa. This tunnel leads to the twisting northern segment of Tecopa and through the aptly named Fallen Stalactite and Hollow Floor Rooms, eventually bringing you into the large main chamber of Tecopa. The tour ends on a platform above the dusty floor in the main chamber of Tecopa, where the ranger will tell of the previous inhabitants, both human and nonhuman.

VIGNETTE 19

THE SAN ANDREAS FAULT NORTH OF THE SAN GABRIEL MOUNTAINS

The Elephant in the Room

LOS ANGELES AND SAN BERNARDINO COUNTIES

When southern California conversations get around to earthquakes, as they inevitably do, geologists are often asked the question: Is California going to break away and slide beneath the Pacific Ocean during the next *big* earthquake? The standard answer is "No." The San Andreas fault is, however, a major force in the grand scheme of our dynamic planet. It is part of the overall recycling process that renews and refreshes the landscape upon which we humans live—although very slowly. Land on its west side nudges northwestward about 1 inch per year. Had one of your authors built a brick path across the fault while a teenager, that path would be offset several feet by now.

Process and rate do matter, however, because this tectonic force accumulates until it is released in earthquakes that can be disastrous if in populated areas. There weren't many people in California in 1857 when the last significant earthquake, the 1857 M7.9 Fort Tejon earthquake, occurred on the southern San Andreas fault. It caused little damage, and only one fatality was reported. Today, almost anywhere along its approximately 700-mile-long trace from Cape Mendocino in northern California south to the Salton Sea, a magnitude 7.9 shaker would certainly cause significant loss of life and infrastructure due to the enormous population increase in California since 1857.

The San Andreas fault is perhaps the best known and most notorious geologic structure in the world. The fault's destructive potential was learned firsthand from the catastrophic 1906 M7.9 San Francisco earthquake and consequent conflagration. Sadly, lessons learned from the earthquake will probably be relearned in heavily populated southern California only on the day the "Big One" occurs along the southern segment of the fault, as it will inevitably.

The 1906 San Francisco earthquake produced a 150-mile-long ground rupture and widespread shaking damage in northern California. Prior to the earthquake, American geologist Harold W. Fairbanks

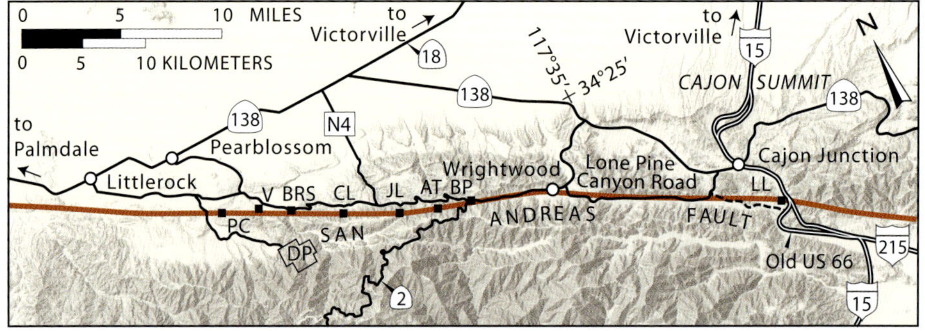

Route map along the San Andreas fault (red line) between Littlerock and Cajon Pass. PC = Pallett Creek; V = Valyermo; DP = Devils Punchbowl Natural Area; BRS = Big Rock Springs; CL = Caldwell Lake; JL = Jackson Lake; AT = Apple Tree Campground; BP = Big Pines; LL = Lost Lake.

GETTING THERE

This vignette follows well-maintained gravel and paved roads along the San Andreas fault. To get to Lost Lake, a sag pond at the beginning of our excursion, drive north on I-15 from San Bernardino to Kenwood Avenue (exit 124). Turn left at the top of the off-ramp and drive west about 0.25 mile under the freeway to the intersection with old US 66. Turn right and follow the old highway north about 2.5 miles. Cajon Creek and the railroad are on your left. The road makes a broad sweeping right turn into the Blue Cut, which takes its name from dark, gray-blue Pelona Schist exposed in roadcuts and railroad cuts on the southwest side of the San Andreas fault. Route 66 follows Cajon Creek out of the cut and crosses the San Andreas fault onto the North American Plate, where the road makes a broad, sweeping curve back to the north. The Z-shaped course of Cajon Creek where it follows old Route 66 results from incremental right-lateral displacement of about 1.5 miles along the fault over many thousands of years (see Vignette 11).

Turn left off old US 66 onto Swarthout Canyon Road, about 1 mile from the Blue Cut, and follow it about 0.4 mile to the first set of railroad tracks. STOP there and look carefully (twice in both directions) for train traffic. The railroads in Cajon Pass are some of the most heavily traveled in the world, and the trains are very quiet. About 0.25 mile beyond the railroad crossing, you cross the San Andreas fault back onto the Pacific Plate where some houses on a ranch on your right (north) are surrounded by tall trees. Continue on the Pacific Plate another 0.25 mile to the next set of railroad tracks. Be careful before crossing here, too. Lost Lake and a fenced parking area are 0.4 mile beyond the second set of tracks.

mapped 400 miles of the fault from north of San Francisco to San Bernardino, directing attention to nearly 300 more miles of its course that had not been explored previously.

The fault might have been recognized fifty years earlier had seismologists been around at the time to study the 1857 Fort Tejon earthquake. Its surface rupture in southern California extended 150 miles from Parkfield in central California southeast to Wrightwood. Maximum displacement along the ground rupture was 30 feet (!) near the junction of the San Andreas and Garlock faults, about 80 miles northwest of Wrightwood. Imagine standing on one side of the fault and suddenly seeing everything on the other side move 30 feet to the right in a few seconds!

The San Andreas fault is one of several subparallel faults in the San Andreas fault system that accommodates horizontal, right-lateral displacement between the Pacific and North American Plates. In southern California, those faults include the San Jacinto, Elsinore, Newport–Inglewood–Rose Canyon, and several offshore faults. Instead of being a single planar break in the upper 15 miles or so of the Earth's crust, each of these large faults constitutes a fault zone as much as several hundred feet wide. Together, they make up the San Andreas fault system, a 600-mile-wide plexus of strike-slip faults, stretching from the northern California coast east into the Basin and Range province of eastern California and Nevada.

Distinctive landforms created during the 1857 earthquake can still be seen along a scenic 25-mile stretch of county roads between I-15 in Cajon Pass north of San Bernardino and Valyermo on the south edge of the Mojave Desert. Let's explore these features, beginning at Lost Lake.

Lone Pine Canyon to Wrightwood

The San Andreas fault is traceable through Lone Pine Canyon by a series of long, low, linear scarps, elongate ponds, and isolated strings of springs. Lost Lake might be better named Hidden Pond because a long low ridge blocks its visibility from Swarthout Canyon Road. From the Lost Lake parking area on that road, walk a couple hundred feet north and up to the crest of the ridge where you can see a classic sag pond—an elongate lake that lies in a long trench-like slot between two opposing fault scarps. A short jaunt down to Lost Lake itself will provide you with a handshake acquaintance with the San Andreas fault.

Lone Pine Canyon's linearity is the result of stream erosion of the crushed rock along the San Andreas fault. If lighting is favorable, you will be able to see several fault scarps paralleling the road along the

Aerial view looking northwest of Lost Lake (34°16.38N, 117°27.92W), a sag pond within an elongate graben between two long fault scarps that, together with a stream terrace scarp on the right, comprise a zone of faults that extends 10 miles northwestward up Lone Pine Canyon to Wrightwood.

southwest side of the canyon. The north ends of some of the ridgelines that descend into Lone Pine Canyon from the hills on the south are trimmed off by the San Andreas fault to form triangular facets, an indicator of faulting.

The San Andreas fault also cuts across tributary drainages that discharge from these hills into Lone Pine Canyon. In map view, these drainage channels are offset with a right-lateral zigzag where they cross the fault.

A few ranches depend on springs at the mouths of these side canyons, where percolation of groundwater is interrupted by the presence of impermeable gouge in the subsurface along the fault. Groundwater typically accumulates on the upslope side of a fault and comes to the surface at the fault. Water-loving vegetation, such as willow trees and cattails, commonly grows in distinctive lines along faults. A good way to locate the fault is to plot these ranches, springs, and vegetation lines on a map.

From Lost Lake, drive about 3.3 miles northwest on Swarthout Canyon Road to where it joins Lone Pine Canyon Road. From this

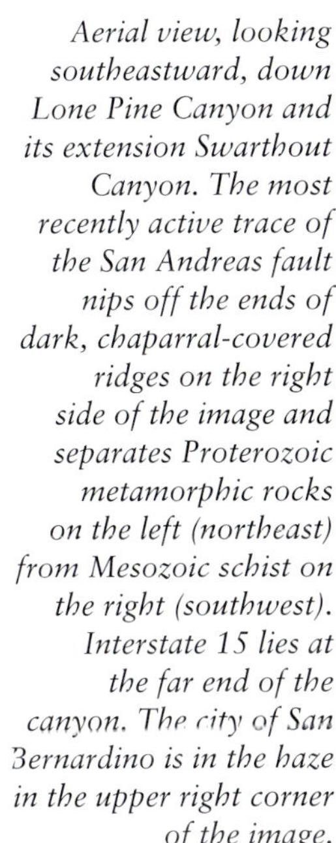

Aerial view, looking southeastward, down Lone Pine Canyon and its extension Swarthout Canyon. The most recently active trace of the San Andreas fault nips off the ends of dark, chaparral-covered ridges on the right side of the image and separates Proterozoic metamorphic rocks on the left (northeast) from Mesozoic schist on the right (southwest). Interstate 15 lies at the far end of the canyon. The city of San Bernardino is in the haze in the upper right corner of the image.

intersection, travel northwest an additional 6 miles to an overlook (34°20.50N, 117°35.98W) on the right (north) side of the road to obtain a splendid view down Lone Pine Canyon. The turnout to the vista point is marked by a line of huge, white-painted boulders on the right (north) side of a gravel road that leads to a very serious white gate, which prevents you from driving farther on an unimproved dirt track to the vista point. Park off the paved road, walk around the gate, and go about a quarter mile southeast to the end of the dirt road. From there, you have a view down Lone Pine Canyon, and you can see that the most recently active trace of the San Andreas fault lies mostly on the right (southwest) side of Lone Pine Canyon.

The fault juxtaposes Proterozoic gneiss and white Paleozoic marble on the northeast side of the canyon against Mesozoic Pelona Schist on the southwest. Rocks along the fault have been so ground up by faulting that they are easily eroded, resulting in an unusually straight canyon, which is quite unlike the meandering nature of most creeks and rivers.

On a clear day, you may be able to see the two highest peaks in southern California on the southeastern skyline: San Gorgonio Mountain (11,499 feet), and on the right, San Jacinto Peak (10,834 feet). The San Andreas fault passes between them through San Gorgonio Pass into the Salton Trough, disappearing at Bombay Beach on the northeast side of the Salton Sea.

Leave the large white rocks and gated entrance to the overlook and continue to drive about 1 mile northwestward on Lone Pine Canyon Road. You'll come to a wide diversion channel two blocks south of Sheep Creek Drive.

Wrightwood is built on, and subject to, mud and debris flows that originate in the rugged San Gabriel Mountains that stand high above the town. The most recent devastating event occurred in May 1941, when rapid snowmelt fluidized a mass of shattered rock and soil in Heath Canyon above town. Debris flows with the consistency of wet cement flowed through the town every day for a week. Rocks and

Gray Pelona Schist detritus lines the Sheep Creek debris flow diversion channel at Wrightwood. (34°20.93N, 117° 36.87W)

mud, deposited upon prehistoric debris flows, covered over 190 acres of Wrightwood, destroying many homes. The flows continued for 15 miles out onto the Mojave Desert floor, where they are prominent on high altitude imagery as a gray alluvial fan on the tan desert floor. Depending on slope and viscosity, an active debris flow may be as deep as 20 feet and travel as fast as 20 miles per hour. Few, if any, vacation cabins would ever survive such an onslaught.

It appears that Wrightwood has invested a lot of faith in diversion channels that have been built through the town. As you drive on through Wrightwood on Sheep Creek Drive (or other nearby residential streets) en route to CA 2, notice that many homes are built on slabs of Pelona Schist left behind by debris flows.

As if the constant threat of debris flows were not enough, Wrightwood is also host to the active San Andreas fault along the southern, upslope edge of town. The 1857 earthquake rupture cut through

Roadcut exposure of a blue gray Pelona Schist debris flow deposit in Wrightwood. Imagine this rock and mud mass coming at you at 20 miles per hour! (34°20.95N, 117°36.908W)

COLLECTING ACTINOLITE

If you have time and an interest in mineral collecting, park off the road near one of the debris channels and take a walk onto the channel surface. Careful poking around will likely turn up some fine radiating clusters of actinolite, a grass-green amphibole mineral. With a little luck you might find patches of fuchsite, a beautiful, bright-green, chrome-bearing mica. Both of these minerals are present in the Pelona Schist.

Wrightwood long before modern development. Several mature pine trees grew right on the fault trace back then. The fault rupture tore up their root systems so severely that the trees' growth was stunted for several years, as confirmed by tree-ring studies. These same trees were also shaken with such force that their tops snapped off as they were whipped violently back and forth. You may spot a couple of these big old trees by the way their trunks rise to a fork about halfway up the tree. Younger split trees were likely damaged by lightning.

Detailed geologic studies in the Wrightwood area have found evidence that more than twenty-three major ground-rupturing earthquakes have occurred here. The range of time between major earthquakes has varied from a low of 31 years to a high of about 165 years, with an average time interval of about 105 years. As of this writing, it has been 163 years since the 1857 earthquake. Take a moment to ponder how Clarence Allen, formerly a Caltech professor of geology and geophysics, used to say that the recurrence frequency of earthquakes on a given fault was equal to the elapsed time since its last earthquake.

Wrightwood will not fare well when the next large earthquake occurs on this segment of the San Andreas fault. Most southern Californians are waiting for the Big One to happen within their lifetimes, but owners of the fifty or more houses on the fault trace hope that it will hold off for many more years.

Turn right (northeast) from Lone Pine Canyon Road onto Sheep Creek Drive and continue 0.5 mile to the T intersection with CA 2. Turn left (northwest) onto CA 2 and follow it northwestward past the town and up the canyon developed along the fault. The most recently active trace of the San Andreas fault lies in the wash along the southwest (left) side of the road. A mile or so west of Wrightwood, the fault lies along the base of the Mountain High ski slopes.

Big Pines to Big Rock Creek

The San Andreas fault reaches its highest elevation anywhere along its entire 700-mile-long length at the Big Pines Ranger Station, 6,865 feet above sea level (34°22.74N, 117°41.39W), 4 miles northwest of Wrightwood. The impressive lodge and US Forest Service visitor center at Big Pines were built to entice the 1932 Winter Olympic Games to the area. Big Pines and Wrightwood bid for the games, but they were awarded to Lake Placid, New York. The visitor center is built upon a low fault scarp that probably formed in the 1857 earthquake. The tower and some of the walls are made of Proterozoic black gabbro quarried in the nearby San Gabriel Mountains.

The 4-foot-high retaining wall at Big Pines Ranger Station consists mostly of angular gabbro cobbles derived from nearby outcrops. (34°22.722N, 117°41.353W)

Leave CA 2 at Big Pines and bear right onto the Big Pines Highway (County Road N-4). Follow this narrow, winding road northwestward down the canyon cut along the San Andreas fault. About 2 miles northwest of Big Pines, look for roadcuts in white, chalky-looking, crushed rock on the north side of the road and for elongate fins of gray gouge in the canyon on the south side of the road. Drive around a long hairpin turn and stop at Apple Tree Campground (34°23.14N, 117°42.4W). From the gate at the east end of the parking lot, you may walk or drive about a quarter mile to the end of a gravel road to obtain an up-close-and-personal view of fault gouge. Intensely crushed rock forms where hard rocks on both sides of the fault have

Fin-shaped erosional remnant of crushed rock (gouge) in the San Andreas fault zone near Apple Tree Campground. (34°23.1'N, 117°42.3'W)

ground one other. At this location, the San Andreas fault juxtaposes Proterozoic gneiss against Mesozoic Pelona Schist.

The topography between Apple Tree Campground and Big Rock Creek is typical of a strike-slip fault. The many elongate ridges, swales, and flats are the result of tens of thousands of years of horizontal displacement and erosion along the fault.

About 0.75 mile west of Apple Tree Campground, spend a few pastoral moments at the elongate Jackson Lake sag pond before you continue onto the southern edge of the Mojave Desert. Jackson Lake lies behind a long, linear shutter ridge (on the north side of the lake). A shutter ridge forms where relatively high ground on one side of a strike-slip fault—for example, a ridge—slides alongside relatively low ground, such as a streambed, on the opposite side of the fault. This ridge consists of sedimentary deposits of the late Pliocene to Pleistocene-age Shoemaker Gravel. These elevated gravels form a natural dam to drainages debouching from canyons on the south side of the lake. Jackson Lake formed naturally in the low area but has been modified and enlarged for recreational use over the years. Crushed basement rock exposed in the roadcut on the north side of the lake is separated by the fault from the Shoemaker Gravel.

Aerial view looking southeast at Jackson Lake, which is trapped behind a shutter ridge of Shoemaker Gravel beds on the northeast side of the road (left). The gravels were uplifted and moved here by right-lateral displacement along the San Andreas fault. (34°23.52N, 117°43.55W)

Shoemaker Gravel deposits are well exposed in hairpin-turn roadcuts along and northeast of the fault zone between Jackson Lake and Big Rock Creek. Distinctive clasts in these gravels were derived from the Pallett Creek drainage, a few more miles down the road and on the other (south) side of the San Andreas fault. This relationship indicates that the Shoemaker Gravel has slipped about 8 miles right-laterally from its source since it was deposited 2 million years ago. Caldwell Lake, another elongate sag pond along the San Andreas fault, is on private fenced property approximately 4 miles northwest of Jackson Lake.

The south-facing fault scarp north of the junction where Big Pines Highway meets Big Rock Creek Road (34°25.968N, 117°50.06W) formed in the 1857 earthquake. We know that because William P. Blake and his fellow railroad surveyors camped here in 1852 and mapped the creek through what is now a low-lying area of agricultural fields east of the scarp. The 1857 earthquake pushed up a scarp

Aerial view looking northwest at Caldwell Lake, a long, narrow sag pond in an undrained depression along the San Andreas fault, 4 miles northwest of Jackson Lake. Evaporation leaves white salt deposits behind. (34°24.86N, 117°46.74W)

that diverted the creek from where Blake mapped it to its present position west of the scarp.

Big Pines Road (County Road N-4) ends at Big Rock Creek. Here the paved road splits into Bob's Gap Road leading right (north), and Valyermo Road that leads left (northwest), subparalleling the San Andreas fault. Continue northwest on Valyermo Road.

Pallett Creek

One of the first places where prehistoric slip history of a fault was determined was in trench excavations at Pallett Creek. Nowadays, it is routine practice to excavate a trench across a fault to determine its slip history, but it wasn't so back in the 1970s when Caltech graduate student Kerry Sieh was the first to develop the procedure.

To get to this hallowed site, drive about 2.5 miles north along Valyermo Road from its junction with Big Pines Road to its intersection with Fort Tejon Road. (See the Getting There map for Vignette 20.) Turn left (southwest) onto Fort Tejon Road. Within less than 0.1 mile

Fort Tejon Road makes a sharp right turn. Stay left (straight) and continue southwest on Pallett Creek Road past St. Andrews Abbey (on your left, south). Go about 1 mile and park in the wide area along the south side of the road near a white post mile marker with a red top and red-lettered 0.76 (34°27.401N, 117°53.25W).

Between the road and Pallett Creek, the San Andreas fault is overlain by stream-deposited silt, sand, and gravel interlayered with peat beds. Sieh's detailed studies of trench exposures revealed that these thin sedimentary layers were offset along several splays of the main San Andreas fault by the 1857 earthquake and earlier ruptures. By radiocarbon dating of peat beds in the trench, he was able to determine when prehistoric fault displacements occurred.

Although Sieh's trench has degraded since the 1970s, it is well worth a walk down the faint path from the parking spot to view exposures in the north bank of Pallett Creek. There, in the sidewall of the trench, you can still safely observe faulted dark-gray peat beds

Alternating beds of black peat and gray pebbly sandstone are broken by a minor fault splay (between arrows) in the Pallett Creek trench. The lower beds are truncated by the fault splay, and the uppermost beds are not. Pen is 7 inches long (to right of upper arrow). (34°27.35N, 117°53.25W)

The trench exposure at Pallett Creek displays black peat beds interlayered with sand and gravel beds.

more than 2,000 years old interbedded with layers of light-gray gravel, sand, and silt. Look low in the trench wall for fault strands that displace beds vertically as much as 6 inches. Follow a fault strand upward through the succession of flat-lying beds to where it terminates. The youngest offset peat bed and the first overlying peat layer that isn't cut bracket the age of slip on that fault strand. Sieh used this procedure for all faults exposed in his trench to determine that nine large, 1857-size earthquakes have occurred at the site since AD 500.

A Forward Look in Hindsight

Except for about fifty homes in Wrightwood, you might conclude that another 1857-style earthquake would do little damage along the route you have just traveled. You'd be correct if you considered only surface rupturing, but severe seismic shaking damage would occur up to 100 miles away from the fault. It will not be a pretty picture in southern California the day that the Big One ruptures water, gas, power, and transportation routes and shakes thousands of square miles of densely populated urban areas and the people in them. Tom Heaton, Caltech professor of engineering seismology, stated ominously, "The San Andreas fault is locked, loaded, and ready to rumble."

VIGNETTE 20

DEVILS PUNCHBOWL NATURAL AREA

A Noble Fold Caught between Faults

LOS ANGELES COUNTY

Devils Punchbowl Natural Area is a remarkable, scenically beautiful county park nearly 5,000 feet above sea level, where you may take a picnic and drinking water for a day of hiking among some wonderfully colorful rocks and geologic structures. The area is a favorite of southern California geology students owing to the inviting swimming holes along some of the creeks, where students may contemplate geology on warm days.

Aerial view looking northwest down the axis of the Punchbowl syncline, which is bounded on the left by the Punchbowl fault trace (between arrows). Pale pink sandstone beds (center) are the Punchbowl Formation; brown shale and sandstone beds (lower) are the San Francisquito Formation. The white crush zone of the Punchbowl fault is left of the Punchbowl Formation. The gray rocks along the left edge of the image are Precambrian gneiss and granite southwest of the Punchbowl fault.

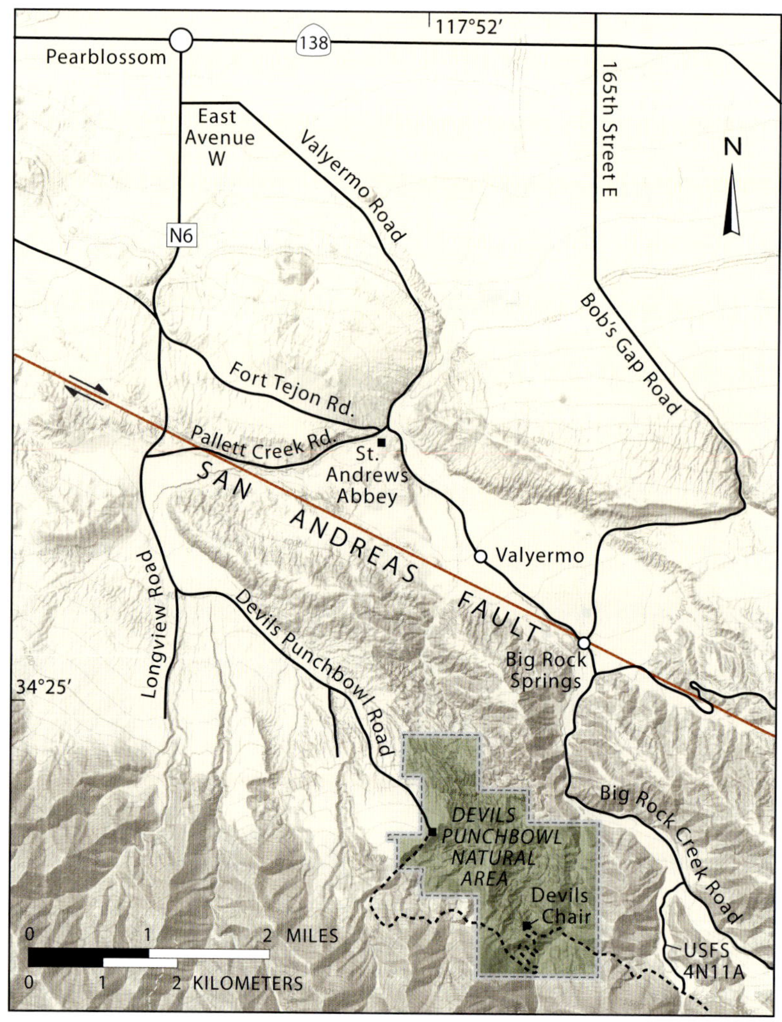

GETTING THERE

Devils Punchbowl Natural Area is in the hills between the Mojave Desert and the San Gabriel Mountains. From CA 14 in Palmdale, drive 20 miles southeast on CA 138 to Pearblossom. Turn right (south) onto Longview Road (N6), proceed 2.25 miles south to Fort Tejon Road; turn left, proceed a quarter mile, then turn right to return to Longview Road. Proceed 2.25 miles south to the left turn onto Devils Punchbowl Road (N6) that leads 3 miles to the Devils Punchbowl Natural Area parking lot (34°24.84N 117°51.5W).

The main geologic feature of the area is a V-shaped, northwest-plunging syncline, known as the Devils Punchbowl. The fold is part of a slice of Earth's crust between the San Gabriel Mountains to the southwest and the Mojave Desert to the northeast. Structurally, the Punchbowl fault zone separates the Punchbowl syncline from Precambrian gneiss and Mesozoic granite of the San Gabriel Mountains, whereas the San Andreas fault juxtaposes the syncline against granitic basement of the Mojave Desert.

The formation of the Devils Punchbowl involved several episodes of sedimentation, folding, uplift, and erosion that began with the deposition of alternating thin layers of sand and silt upon Mesozoic granite in a shallow sea about 60 to 40 million years ago during

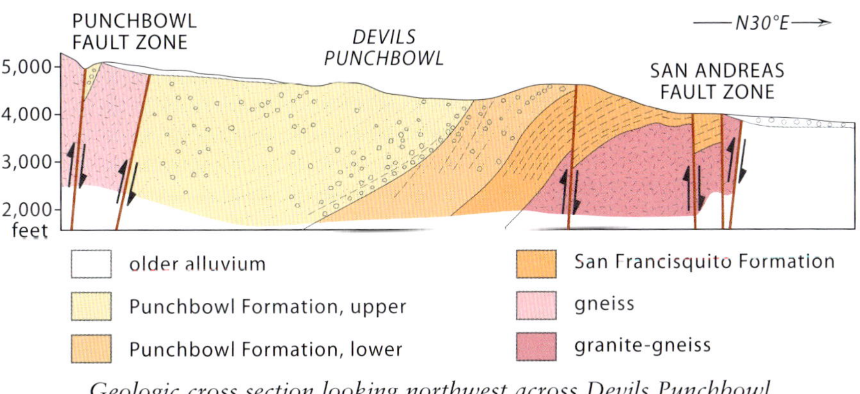

Geologic cross section looking northwest across Devils Punchbowl between the Punchbowl and San Andreas faults. —Dibblee, 1987

AGE	ROCK UNIT	THICKNESS	DESCRIPTION
QUATERNARY	alluvial fan (2.6 million years to present)	0 to 50 feet	sand and gravel
LATE MIOCENE	Punchbowl Formation (10 to 5 million years)	5,000 feet	nonmarine conglomerate, sandstone
PALEOCENE EOCENE	San Francisquito Formation	4,700 feet	marine shale and sandstone
MESOZOIC	Pinyon Ridge		granodiorite PUNCHBOWL FAULT
PROTEROZOIC	San Gabriel Gneiss (more than 1,700 million years old)		gneiss

Rock units at Devils Punchbowl. Wiggly lines are unconformable contacts; solid black line is the Punchbowl fault. Note that clasts of the Lowe Granodiorite are abundant in streambeds, but the rock unit does not crop out here.

Alternating sandstone and shale beds of the San Francisquito Formation are steeply tilted in the Punchbowl syncline. Staff is 4.5 feet long. (34°23.92N, 117°49.75W)

Tilted pebbly sandstone layers of the Punchbowl Formation are in the foreground, and the snowcapped San Gabriel Mountains are on the skyline. (34°24.902N, 117°51.381W)

Paleocene and Eocene time. These sediments were buried and hardened into buff sandstone and dark micaceous shale, the San Francisquito Formation.

Sandy and conglomeratic sediments eroded from a terrane to the north were deposited upon an erosion surface of folded and faulted San Francisquito strata between 10 and 5 million years ago and then hardened into 5,000 feet of thick, buff beds of sandstone and conglomerate called the Punchbowl Formation. It contains mammalian fossils including horses, camels, hippos, antelopes, and dogs.

Both formations were folded into the Punchbowl syncline and uplifted in late Pliocene time, about 3 million years ago. Half of the syncline was cut by and displaced far to the northwest along the Punchbowl fault; the other half is here in the Devils Punchbowl Natural Area. Renewed uplift and erosion occurred in Pleistocene time, followed by deposition of a thin veneer of sand and gravel upon an erosionally beveled surface of low relief across the San Francisquito and Punchbowl strata. Today, creek erosion is cutting deep canyons into all of these rock formations and exposing the subsurface structure.

Much of the Punchbowl Formation has been eroded away, although the Punchbowl rocks are remarkably similar to the 8,000 feet of north-dipping strata at Mormon Rocks, about 25 miles southeast of Devils Punchbowl. The two similar exposures are on opposite sides of the San Andreas fault, and their separation was first cited by geologist Levi Noble as evidence for large-scale right-slip on the fault. The correlation of the two localities is more apparent than real, however, underscoring how geologic interpretations based on limited data may change as new observations are made. First, fossil mammals collected from the two localities many years after Noble's work indicate the rock layers at Devils Punchbowl are several million years younger than those at Mormon Rocks. Moreover, Devils Punchbowl sandstone contains granite and volcanic clasts, including distinctive clasts of a polka-dot granite. This rock is lacking at Mormon Rocks; in its place is phyllite, a flaky metamorphic rock that is not present in the Devils Punchbowl rocks. Geologists have concluded that the rocks in the two areas were derived from two different sources at different times, are not correlative, and thus cannot be used to determine fault displacement.

The San Andreas fault system became active during late Miocene and early Pliocene time. The Punchbowl fault, now inactive, was an ancestral strand of the San Andreas fault, originating about 5 million years ago after deposition of the Punchbowl Formation.

It accumulated about 25 to 30 miles of right slip before it became inactive in Pleistocene time when the plate boundary shifted a couple miles northeastward to the present position of the San Andreas fault. When that happened, the Punchbowl slice, including the Punchbowl syncline, was transferred from the North American Plate to the Pacific Plate, and is now headed for Alaska.

The Punchbowl fault, which dips 70 degrees south in its surface exposures, has both reverse slip and right-lateral strike-slip components of displacement. Shearing along the fault has reduced the Precambrian gneiss and granite to a quarter-mile-wide zone of severely crushed rock between the ancient Precambrian rocks of the San Gabriel Mountains to the southwest and the younger San Francisquito and Punchbowl Formations to the northeast. These rocks are beautifully exposed along the trail to the Devils Chair.

Exploring the Punchbowl

Start your visit by stopping by the visitor center (34°24.844N, 117°51.490W) and then walk over to the adjacent Punchbowl overlook to view the Punchbowl syncline. Notice how the tilted, slab-like layers of thick, buff sandstone beds on each side of the deep canyon are inclined steeply toward each other, defining the syncline. The pebbly sandstone beds are more resistant to erosion and form the slabs, whereas the interbedded, finer-grained sandstone and silty sandstone erode more easily and are vegetated. A 1-mile-long loop trail starts near the overlook and takes you down and around these rocks for a close inspection of them.

A more scenic but rigorous geologic hike takes you to the Devils Chair (34°24.149N, 117°50.747W). Before embarking on this hike, however, have a look at the metamorphic and igneous rocks in the low wall at the east end of the visitor center parking lot. They represent the kinds and types that make up most of the San Gabriel Mountains. The gray-and-white banded rocks are Precambrian biotite gneiss; even-grained salt-and-pepper rocks are Mesozoic granodiorite. Look especially for distinctive rocks with greenish-black spots resembling the speckled back of a trout. The spots are hornblende crystals in an igneous rock of Permian to Triassic age known as the Lowe Granodiorite; it takes its name from widespread exposures on Mt. Lowe in the central San Gabriel Mountains. Boulders of all these rocks are prevalent in streams and creeks issuing northward from the San Gabriel Mountains onto the Mojave Desert, especially in Big Rock Creek along the southeast edge of the Punchbowl. Early-day

View looking southeast of the Punchbowl syncline as seen from the visitor center. Symbols indicate the strike and dip of the sandstone beds. (34°24.8N, 117°51.1W)

Crushed white basement rocks are juxtaposed against pink Punchbowl Formation rocks by the Punchbowl fault (34°24.176N, 117°50.77W), looking northwestward from Devils Chair (34°24.149N, 117°50.747W).
—Photo by Fred Chester

LEVI NOBLE PIONEERING GEOLOGIST

Levi Noble at his Valyermo ranchhouse in 1959.

Not far from Devils Punchbowl Natural Area and located directly athwart the San Andreas fault is Valyermo Ranch, the former long-time residence of Levi Noble (1882–1965), a laconic New Englander and legendary US Geological Survey (USGS) geologist. As a Yale University graduate student from 1908 to 1909, Noble was one of the first geologists to study the geology of the Grand Canyon after John Wesley Powell's pioneering exploration of the canyon in 1869. Imagine the thrill of being among the first scientists to study the stratigraphic succession of rocks in the Grand Canyon, naming many of the sedimentary formations there, and making the first detailed geologic map of part of the canyon.

In 1910, Noble married Dorothy Evans, whose parents gave the ranch to them as a wedding present. According to the Nobles, they were given their choice among that ranch, or a large parcel of land in Los Angeles near what is now the intersection of Hollywood Boulevard and Vine Street, or a hill in the Long Beach area that later became a billion-barrel oil field. Levi and Dorothy chose the ranch on the San Andreas fault because it provided ready access to study that world-famous geologic feature. The ranch was their principal residence for the rest of their lives.

Once ensconced on the ranch, Levi worked intermittently on the San Andreas fault from 1909 to 1954. In 1940 he commenced fieldwork for the geologic map of the Valyermo quadrangle, which includes the Devils Punchbowl area. By the late 1940s or early 1950s, however, the USGS thought it was high time he published his San Andreas maps. According to folk legend, the USGS sent Phillip B. King, distinguished member of the survey, to aid Levi at the ranch and keep him focused. While Levi mapped and wrote, King contributed cross sections for Levi's map of the Pearland 7.5' quadrangle and sketched scenes that were published with Levi's map of the Valyermo 7.5' quadrangle. These maps are available from the US Geological Survey.

Geologic studies by Noble and other notable geologists, including Mason Hill, Thomas Dibblee, Vincent Matthews III, and John Crowell, led to recognition of the remarkably large lateral displacements along the major faults of the San Andreas system in southern California, which shortly thereafter became understood in the context of the then-developing theory of plate tectonics.

settlers used them for walls and building foundations. Outcrops of these same rocks are crushed beyond all recognition where you'll encounter them along the Punchbowl fault on the trail to the Devils Chair.

The 3.7-mile-long trail (7.4 miles round-trip) to the Devils Chair is easy to follow all year unless snow-covered in winter months. Drinking water is not available at the park or along the trail, so take plenty whatever the season. The trailhead is at the southeast corner of the parking lot. Once on the trail, keep left at the green shed and continue for about 2 miles. At a midway point along the trail, you will cross a very thin slice of dark-red pebbly sandstone in the lower part of the Punchbowl Formation caught along the southernmost, secondary trace of the Punchbowl fault.

The Devils Chair will come into sight as you traverse southeastward along the ridge. Continue along the trail to the beginning of a series of switchbacks that leads down to a narrow ledge with a metal railing on each side that goes out to the viewpoint. Be careful, watch your step, and enjoy the view, especially to the northwest of the Punchbowl fault juxtaposing crushed white basement rocks and pink layers of Punchbowl Formation, and eastward across the Punchbowl syncline toward the Mojave Desert.

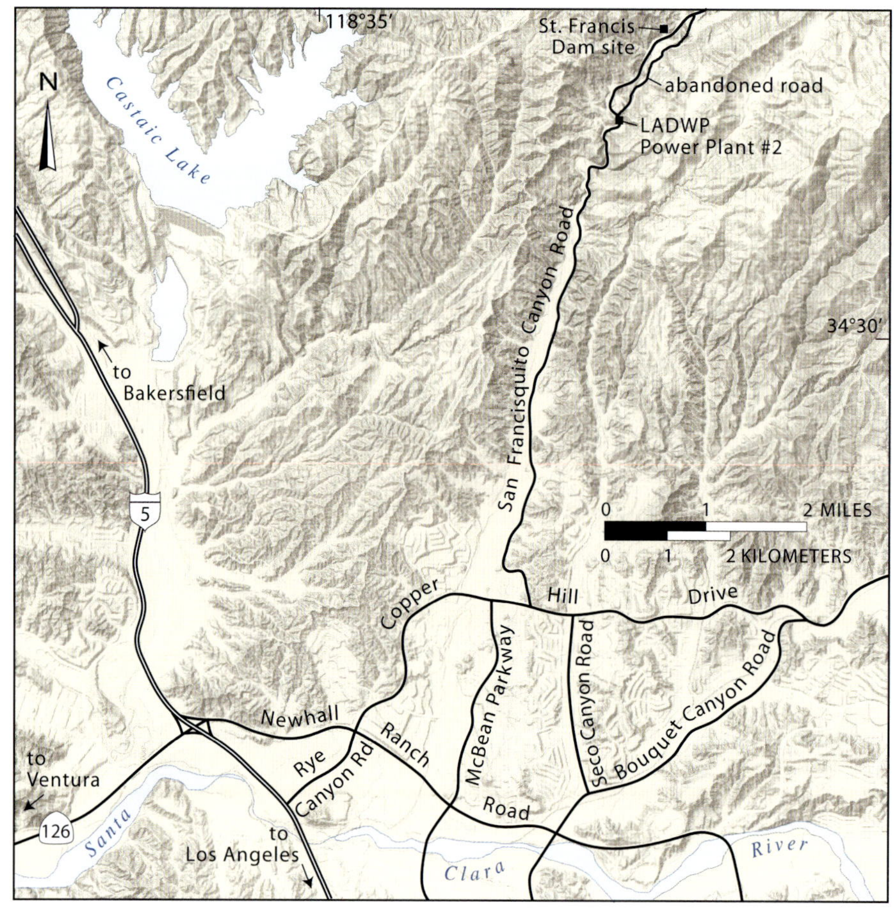

GETTING THERE

From I-5 in Valencia, take Newhall Ranch Road (exit 172) and proceed 1.5 miles east to Copper Hill Drive, where you can turn left and drive 3 miles to San Francisquito Canyon Road. Turn left again and proceed 5.5 miles to Los Angeles Department of Water and Power (LADWP) Power Plant No. 2. A bronze plaque at the north end of and behind the power plant fence commemorates the dam failure. The old road through the dam site is abandoned, but it can be walked or bicycled. Its south end is 0.2 mile beyond the power plant. You may park in front of the concrete K-rail that blocks the road and walk north about 1.3 miles up the old road through the dam debris field to the dam site, or drive another 1.5 miles from the power plant on the main road to the north end of the abandoned road. Park there and walk south about 500 yards south-southwest along the abandoned road to the concrete rubble pile marking the location of the former dam.

VIGNETTE 21

ST. FRANCIS DAM

Disaster at Midnight, 12–13 March 1928

LOS ANGELES COUNTY

Two things keep geotechnical and civil-design engineers awake at night: Dam failures and bridge failures. But fortunately, engineers learn from their mistakes. The St. Francis Dam failure was costly in terms of loss of life and property; it also cost a prominent man his job and reputation, yet many lessons were learned.

Dams should not fail. From the moment of inception to dedication and ribbon cutting, extreme care and caution must be exercised by dedicated teams of scientists and civil-design engineers, as well as contractors and inspectors. All members of the team must be able to communicate with each other in clear, understandable terms, so that in the end, all will sleep well at night.

St. Francis Dam was built in San Francisquito Canyon between 1924 and 1926 by the City of Los Angeles' Bureau of Water Works and Supply to create a large regulating reservoir. Water was to be conveyed 233 miles to the dam from Owens Valley via the Owens Valley Aqueduct. The original aqueduct was completed in 1913 but was extended northward into the Mono Basin in 1940, resulting in a 419-mile-long aqueduct. The reservoir was designed to hold enough water (about 30,000 acre-feet) to supply thirsty customers should the water supply be temporarily disrupted by an earthquake on the San Andreas fault, or by early-twentieth-century water conflicts, as memorialized in the movie *Chinatown*.

The failure of St. Francis Dam is commonly regarded as the worst American civil engineering disaster in the twentieth century. Its catastrophic collapse on March 12, 1928, at 11:57 p.m. sent a 70-foot wall of water down San Francisquito Canyon. A 70-mile-long, 2-mile-wide swath of land was denuded of vegetation as the deluge made its way through the Santa Clara River valley on its way to the Pacific Ocean. More than 450 people perished between Valencia and Oxnard during the 5.5-hour rampage. Boulders, mud, and trees, swept up by 12 billion gallons of floodwater, destroyed 900 homes, many bridges and roads, and 24,000 acres of farmland.

Construction of this concrete gravity arch dam began on March 9, 1925, and was completed a year later—almost to the day. Its twin, Mulholland Dam, was constructed in the hills overlooking Hollywood, across a small canyon on the south side of the Santa Monica Mountains. Located about 30 miles south-southeast of St. Francis Dam, it was completed in December 1924 and is alive, well, and functioning today after some modifications. It, too, was constructed as part of the City of Los Angeles' water supply system.

St. Francis Dam's initial design was deemed suitable for the deep, narrow part of San Francisquito Canyon, which widens upstream into a broad reservoir area. In hindsight, nothing was found to be wrong with the dam's initial design, but many things went wrong with modifications that increased its height, and its geological foundation was later determined to be very poor.

If anyone did witness the dam failure, they did not survive to tell the tale. Thus, determining the sequence of events leading up to and during the actual collapse is educated guesswork. A number of

Panoramic view of St. Francis Dam in 1926. The colossal concrete structure was 205 feet high, 1,225 feet long on its crest, 150 feet long at its base, and 160 feet thick at stream level. —Photo taken by Monolith Cement Company © City of Los Angeles Department of Water and Power

Aerial view looking north of St. Francis Dam site in 2018. (34°32.82N, 118°30.75W)

observations made both before and after the failure bear on what may have actually happened:

1. Cracks in the concrete dam were observed to be leaking water prior to the failure—a sign of inherent concrete weakness and potential dam failure—but the leakage was determined to be within normal limits for a dam of its size and type of construction.
2. The northwest end of the dam abutted in poorly consolidated sandstone and conglomerate of the Vasquez Formation in fault contact with Pelona Schist—a prescription for the percolation of water under the dam through the cracks in the rocks and along the fault contact.
3. The southeast end of the dam abutted into a dip slope of relatively weak, landslide-prone Pelona Schist—a recipe for an unstable foundation.
4. Prehistoric landslides occurred in the Pelona Schist on the southeast side of San Francisquito Canyon, both up- and downstream from the dam. The dam site was selected partly because the canyon was narrow there. It was not recognized in the 1920s that

the narrowing of the canyon was due to an ancient mega landslide at that location.

5. A new landslide of Pelona Schist into the reservoir pool in the hours prior to the dam failure may have created a wave that could have overwhelmed the already cracked concrete dam. An event like this would have tipped the unstable dam system into failure and disaster.
6. It is unclear whether or not loose alluvium was removed from the stream channel at the base of the dam itself. Poorly consolidated older alluvium was apparently left in place under the right wing wall, the concrete wall that extended laterally from the northwest side of the main dam.

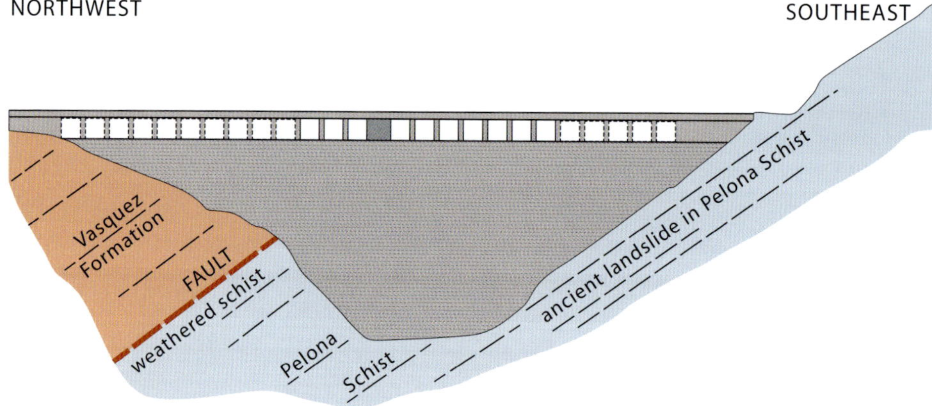

Schematic elevation view of St. Francis Dam and its geologic foundation looking upstream before its height was raised 10 feet and before a 200-foot-long wing wall, 8 feet high, was built across the ridge at the northwest end of the dam.

You can wander the site and examine the rocks and geologic factors that made it unsuitable for a dam—at least for the type of dam that was built—as well as the remnants of the failed dam itself. Most of these features can be seen from the abandoned road in the canyon bottom. A splendid overall view of the site may be gained by climbing to the top of an unnamed ridge (34°32.831N, 118°30.872W) via an unmarked trail that takes off behind the big oak tree by the service road gate (34°32.760N, 118°30.960W) on the south side of the ridge. This viewpoint is located approximately 0.4 mile southwest of the dam site. We do not advise crossing the dam site from the abandoned road to the southeast wall of the canyon because you'll either get your feet wet in the swampy stream or awaken rattlesnakes lurking in the unstable, broken concrete blocks. Keep a wary eye open for rattlesnakes everywhere because they are commonly found in this area.

When the dam broke, rushing water carried pieces of the dam more than a half mile downstream. Look downstream from the vista point noted above and see some of the largest concrete remnants of the dam that rise above the canyon floor. One of the largest, about the size of a small house, is located at 34°32.793N, 118°31.074W.

At the dam site, two rock formations are juxtaposed by the San Francisquito fault, which strikes along the edge of the abandoned road. Red-brown Vasquez Formation of Oligocene age is on the northwest side of the fault, and gray Pelona Schist of Mesozoic age is on the southeast side. The fault is considered inactive because it does not cut 10-million-year-old late Miocene strata a few miles downstream. Many dams worldwide are built across faults, both active and inactive, without unusual problems, so faults, by themselves, do not preclude dam construction across them.

Downstream view of San Francisquito Canyon below the St. Francis Dam site. Arrows point to two remnant piles of the failed dam that were carried about 300 yards downstream by the deluge (right center). Another remnant pile is circled where the abandoned road bends left out of sight.

The Vasquez Formation is exposed in roadcuts along the abandoned road. This nonmarine sedimentary formation consists of poorly consolidated, gypsum-cemented sandstone and pebbly conglomerate that dip southwestward—down the canyon.

This downstream bedrock dip is bad for dam stability. In this configuration, reservoir water can easily enter the up-dip ends of the sandy strata and percolate down-dip, past the dam, to discharge beyond the dam. Over time, water seeping through poorly consolidated sand and along fractures will erode larger and larger passageways. This process, known as piping, can lead to erosion and failure of the dam's rock foundation.

Worse, however, some of the Vasquez strata are cemented with gypsum, which is slightly soluble in water and upon being wetted can cause the rock to disaggregate, making it easier for piping to occur. Moreover, the rock itself is not especially well consolidated; that is, it is more like a dirt pile rather than good solid foundation rock.

You can examine the schist in the abandoned roadcut about 125 yards northeast of the service road gate. Less fractured, more typical exposures of the schist crop out south of the dam site where the abandoned road crosses the creek near the service road gate (34°32.739N, 118°30.996W).

Metamorphic rocks are generally hard and compact, quite suitable for a dam foundation. Not so the Pelona Schist, however, which is a low-grade metamorphic rock with a sheeted or schistose structure like the pages in a book. In San Francisquito Canyon, the schist breaks readily along planes of schistosity into slabs that dip into the canyon, where they can slide into the reservoir. Look closely at the southeast canyon wall, upstream of the dam site, to see faint outlines of shovel-shaped landslides in the schist that have occurred since the dam collapsed.

ACTIVITY TO TEST ROCK COMPETENCY

Bring a 5-gallon bucket with you for this trip to the St. Francis Dam site. Fill it with creek water when you arrive at the dam site and drop a chunk of Vasquez Formation sandstone into the water. Put a piece or two of Pelona Schist into the bucket, too. Give them a couple of hours and then see how the Vasquez clasts slowly disaggregate as the gypsum dissolves. As an alternative, bring a chunk of the rusty-red rock home and perform the experiment where you can let it sit as long as you want.

Roadside outcrop of the sandstone and pebbly sandstone of the Vasquez Formation in the northwest canyon wall used as a foundation for the St. Francis Dam. Note how the faint bedding layers (dashed lines) dip about 30 degrees downstream, to the left. The yucca stalk and flower are 8 feet tall. (34°32.843N, 118°30.779W)

Roadcut exposure of the gray Pelona Schist cut by white quartz veins. Sunglasses (upper right) for scale. (34°32.786N, 118°30.857W)

You'll also find unconsolidated alluvial sand and gravel at two locations within the dam site. One is where the dam once stood. All of it should have been removed prior to pouring concrete, so that the dam would sit directly on and bond tightly to the underlying bedrock. Complete removal of alluvium, as well as work to seal cracks in the bedrock, is necessary to prevent water from escaping under a dam by means of seepage or piping. Modern alluvium, washed into the former dam site after the dam failed, occupies the stream bottom today.

A thin layer of older alluvium underlies the northwest wing wall at the crest of the Vasquez Formation ridge line. These deposits should have been removed prior to wing-wall construction as well.

Dr. J. David Rogers, a professor of geological engineering at Missouri University of Science and Technology, attributed construction of the southeast end of the dam against a large ancient landslide as one possible cause that led to dam failure. He cited evidence pointing to incipient sliding above the dam in the hours before its collapse.

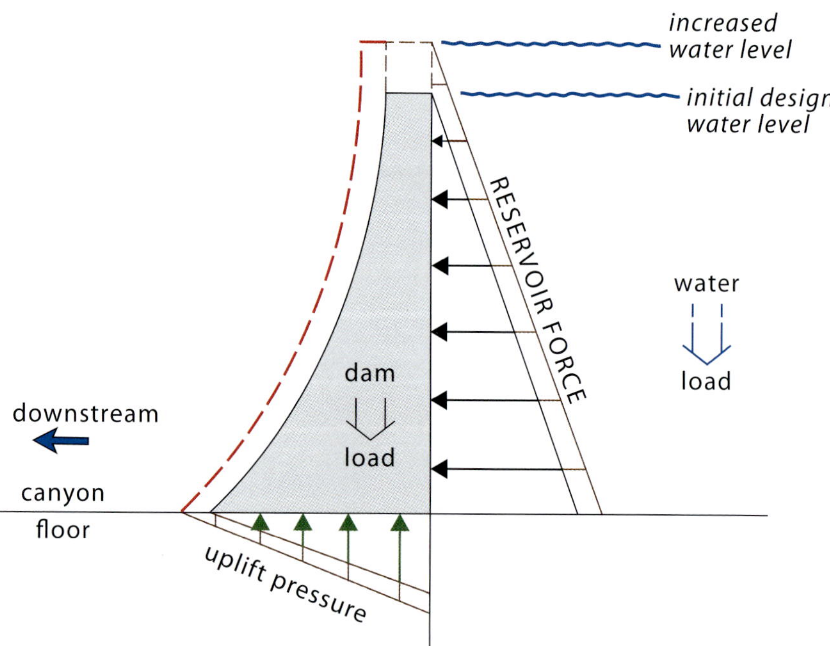

Schematic cross-sectional diagram of a dam and of forces on and beneath a dam whose stability is provided by its load weight. Merely raising the height of the dam to increase the reservoir capacity increases the external reservoir forces (black), water load (blue), and hydraulic uplift pressure (green). These increased forces must be counteracted by increasing the overall size of the dam (dashed red line) and, therefore, its weight to prevent it from tipping or sliding downstream.

Rogers also noted that the dam height was raised by 10 feet and the northwest wing wall was added to increase reservoir volume from 30,000 acre-feet to over 38,000 acre-feet after the final dam design. The size of the dam footprint should have been increased to provide additional support to balance the increased reservoir water pressure against the dam, but this was not done.

Following the disaster, a state commission concluded that the dam failed because it was poorly constructed on a geologically unstable site. The blame for these defects and thus for the dam failure was laid upon, and accepted by, a single individual—William Mulholland, who conceived of and constructed the Los Angeles Aqueduct. Mulholland, a self-taught engineer whose long and distinguished career included construction of the Owens Valley Aqueduct in 1908–1913, was devastated by the failure and retired later in 1928.

Because of the catastrophic St. Francis Dam failure, the California State Legislature updated its dam safety program in 1929 by creating a Division of Safety of Dams (DOSD). This division, under the Department of Water Resources, was tasked with ensuring that all dam designs and specifications receive appropriate peer review and approval prior to moving forward with construction. In addition, a three-member board of independent consultants must be included with the review team to avoid conflicts of interest.

At least once every five years, the independent consulting board reviews all nonfederal dams higher than 25 feet and dams having a reservoir capacity greater than 50 acre-feet. The state also has full authority to supervise maintenance and operation of all nonfederal dams.

Glossary

a'a. A Hawai'ian term for a type of basaltic lava with a blocky, jagged, clinkery, spiny surface.

accretionary wedge. A wedge-shaped mass of sediment and rock fragments that has been scraped off the surface of a downgoing oceanic plate and accumulated where the oceanic plate meets an overriding plate at a subduction zone.

alluvial apron. Smooth deposit of alluvium, regional in extent, sloping gently outward from the base of a mountain face.

alluvial fan. Low, fan-shaped, gently sloping stream deposit of alluvium bordering the base of a steep slope at the mouth of a canyon.

alluvium. Unsorted, unconsolidated boulders, cobbles, gravel, sand, and finer rock debris deposited in relatively recent geologic time, principally by running water; adjective: **alluvial.**

andesite. A dark-colored, fine-grained volcanic rock intermediate in composition between rhyolite and basalt.

angle of repose. Maximum angle of a slope, measured from horizontal, assumed by a pile of loose, cohesionless (dry) rock particles.

angular unconformity. A break or interruption of the geologic record within a sequence of layered rocks, in which younger, upper layers rest upon an erosional surface that angularly truncates older, lower layers.

antecedent stream. Said of a stream or river established before uplift developed across it and that maintains its course across and in spite of the growing uplift by incising its channel at the same rate as the land rises.

anticline. A fold in layered rocks, convex upward, with stratigraphically older rocks in its core.

arroyo. A relatively broad, flat-floored, steep-banked channel or gully of an intermittent stream in arid or semiarid country.

ash. See **volcanic ash.**

ash-flow tuff. A volcanic rock consolidated from an ash flow—a highly heated mixture of volcanic gases and ash that travels as a density current down the flanks of a volcano or along the surface of the ground.

asthenosphere. The ductile, highly viscous upper layer or shell of the Earth below the lithosphere; includes the upper mantle.

axis. As used herein, the center line of a fold in rocks.

badland. Barren, steep, intricately dissected topography developed by erosion in fine-grained but coherent sediments.

basalt. Fine-grained, dark, primarily extrusive igneous rock, relatively rich in calcium, iron, and magnesium, and relatively poor in silicon.

basement. Undifferentiated complex of rocks that underlies the rocks of interest in an area.

beach. A constantly changing deposit of loose sand or stones along the gently sloping shoreline of a water body.

bedding. The layered arrangement and structure of sedimentary rocks.

bedrock. Relatively solid rock, exposed or underlying a mantle of soil or loose rock detritus.

bluff. High bank, cliff, or headland rising above a flat.

boulder. A rock fragment larger than 10 inches in diameter, usually worn by abrasion during transport and at least partly rounded.

brachiopod. Clam-like solitary marine invertebrate whose shell pieces are bilaterally symmetrical.

breakwater. An offshore structure designed to protect an anchorage from the force of waves.

breccia. A rock consisting of angular broken rock fragments held together by a mineral cement or in a fine-grain matrix.

calcite. Widespread, abundant mineral composed of calcium carbonate ($CaCO_3$); the major component of limestone and marble.

carbon-14. The radioactive isotope of carbon, which disintegrates with a half-life of 5,730 years, plus or minus 40 years.

carbonate rock. Rock composed of the minerals calcite or dolomite, both of which contain carbonate (CO_3); typical examples are limestone and marble.

cave. Naturally formed subterranean chamber, large enough for a person to enter.

cavern. A large cave or cave system.

chaparral. Thicket-like vegetative complex of scrubby bushes densely cloaking areas in semiarid regions.

chert. Hard, dense, dull to partly glassy sedimentary rock composed of finely crystalline silica.

cinder cone. Cone-shaped accumulation of volcanic cinders erupted from a central basaltic or andesitic vent; see **volcanic cinders**.

clast. An individual rock fragment, regardless of size, shape, or texture, produced in mechanical weathering and large enough to be visible to the naked eye in sediment or sedimentary rock; adjective: **clastic**.

clay. Rock or mineral particles smaller than 0.00016 inch, or crystals of a clay mineral; plastic when wet.

cleavage. Ability of many minerals to split along crystallographic planes.

cliff. A high precipice in rock, near vertical or overhanging.

coarse-grained. Said of sedimentary rocks that have clasts, or particles, that are relatively large. Said of igneous rocks with relatively large crystals.

cobble. Rock fragment between 3 and 10 inches in diameter, usually worn to rounded by abrasion in the course of transport.

conglomerate. Sedimentary rock consisting of rounded pebbles, cobbles, or boulders, cemented within a sandy or silty matrix.

crinoid. Also known as a sea lily; a bottom-dwelling animal with a slender column supporting a globular, pentagonal body from which appendages extend. Generally, only cylindrical stem segments are preserved.

cross bedding. Arrangement of sedimentary beds inclined at an angle to the main bedding.

crystal. A many-faced solid bounded by smooth planar surfaces that reflect an orderly internal arrangement of atoms.

debris. Surficial accumulation of loose, broken rock and soil fragments.

detritus. Collective term for loose, disaggregated fragments of rock worn off or removed by mechanical abrasion.

dike. Tabular, igneous intrusive body, discordant with structure of country rock.

dip. The inclination from horizontal of any planar surface within rocks, as measured in the steepest direction of that surface (e.g., the direction a marble would roll down the surface).

displacement. A general term for the relative movement of the two sides of a fault, measured in any chosen direction.

dolomite. A common rock-forming carbonate mineral with the formula $CaMg(CO_3)_2$; term is also applied to a sedimentary rock consisting of dolomite.

dripstone. Cave formation, generally of calcite, formed by dripping or seeping water.

earthquake. Shaking within the Earth's crust produced by an abrupt release of accumulated strain.

eddy. A temporary circular movement of water or wind, usually in a different direction from that of the main current.

erosion. The mechanical destruction and removal of rock material by any natural process.

estuary. Seaward end of a shoreline inlet formed by the drowning of a river valley.

fault. Fracture along which blocks of Earth's crust have slipped past each other.

feldspar. Group of common rock-forming minerals composed principally of silicon, aluminum, and oxygen, plus one or more of the elements calcium, sodium, and potassium.

fine-grained. Said of sedimentary rocks that have clasts, or particles, that are relatively small. Said of igneous rocks with relatively small crystals.

flowstone. Typically any deposit of calcium carbonate formed by flowing water on walls or floor of a cave.

foliation. A general term for the planar arrangement of minerals or structural features in rocks, formed primarily by solid-state metamorphism. Such a rock is said to be foliated.

foredune. A coastal sand dune oriented parallel to the shoreline and positioned at the landward margin of a beach.

formation. Geologically, a rock body of considerable areal extent with consistent characteristics that permit it to be recognized, mapped, and usually named.

fusulinid. One-celled, floating marine animal with a cucumber-shaped, chambered shell resembling a grain of wheat up to 1 inch in length.

glacier. A large body of natural, land-borne ice that flows.

gneiss. A rock formed by regional metamorphism with a foliation that is more widely spaced, less marked, and commonly more discontinuous than that of a schistose rock.

gouge. A thin layer of soft, earthy, crushed rock along a fault.

graben. An elongate, relatively depressed crustal block that is bounded by faults on its long sides.

granite. A completely crystalline plutonic rock in which quartz constitutes 10 to 50 percent of the light-colored minerals.

greenstone. A field term for any compact, metamorphosed, iron- and magnesium-rich igneous rock, with greenish minerals, usually hornblende and chlorite.

groin. Narrow jetty of large stones or piling extending out from shore, designed to capture longshore-drifted sand.

groundmass. Fine-grained part of a rock (usually igneous) in which larger particles or crystals are imbedded.

ground motion. A general term for all seismic motion.

groundwater. Water that fills pores and other openings to a condition of saturation in subsurface rocks and sediment.

group. A rock unit next in rank above formation and thus may be subdivided into formations. No rock type designation is included or implied.

ice age. A period in Earth history when large sheets of ice inundated parts of nonpolar continents.

igneous. Said of rocks or minerals formed by the crystallization of molten material (magma).

iron oxide. General term for any one of the distinct natural iron-bearing minerals, including hematite, goethite, magnetite, and siderite.

isotopic dating. Determining the age of a geologic sample by measuring isotopic ratios of parent to daughter products. An **isotope** is a species of an element defined by the number of neutrons in its nucleus.

lagoon. Shallow, shoreline body of seawater separated from the open sea by an offshore bar or reef; adjective: **lagoonal.**

lava. Extruded magma or the solidified product of such.

limestone. Sedimentary rock composed largely of the mineral calcite.

linear dune. Long, narrow sand dune ridge parallel to prevailing wind.

lithosphere. Rigid outer part of the Earth, includes the crust and upper mantle.

longshore drift. Movement of material, mostly sand, parallel to the shore of a sea or lake, owing to the oblique approach of waves.

magma. Naturally occurring molten rock within the Earth and capable of intrusion and extrusion.

magnetite. Black, magnetic mineral composed of oxides of iron.

magnitude. For earthquakes: A measure of strain energy released during abrupt release of stored elastic strain energy along a fault. Magnitude is measured on a logarithmic scale, with each increase of one unit of magnitude corresponding to a ten-fold increase of amplitude of the trace written on a standard seismograph but a thirty-fold increase in the energy released.

mantle. The plastic zone of the Earth between the core and the crust.

marble. Metamorphosed limestone or dolomite, usually coarsely recrystallized.

massive. Said of a rock layer (such as a thick sandstone bed) or a rock type (such as granite) without evident layering.

matrix. Fine-grained rock or mineral particles filling spaces between coarser constituents of a sedimentary rock.

metamorphic rock. A rock that has undergone sufficient solid-state physical changes by heat, pressure, and shearing stress to be distinct from the parent rock.

mica. Group of rock-forming silicate minerals with perfect sheet-like cleavage; includes biotite and muscovite.

microcrystalline. The texture of a rock consisting of crystals too small to be recognized or distinguished under an ordinary microscope.

mineral. Homogeneous, naturally occurring, inorganic, solid substance of fixed chemical composition and physical properties or that vary within specific limits.

mudstone. Fine-grained sedimentary rock of silt and clay, coarser-grained and more massive than shale; indurated mud.

narrows. Topographically, a gorge or constricted passage along a stream or through a pass.

nodular. As used herein, a roughly spherical knot of mineral material within a rock, commonly in a mica-bearing schist.

normal fault. An inclined fault on which the rock mass above the fault has moved relatively downward.

North American Plate. One of eight huge moving plates that make up the outer-solid part of the Earth (lithosphere).

obsidian. Black or dark-colored volcanic glass; usually rhyolite lava that cooled too quickly to form crystals. Characterized by conchoidal fracture.

oreodont. Heavily built, short-legged, pig-like animal with four-toed hoofs.

outcrop. An exposure of bedrock at the Earth's surface. The rock is said to "crop out."

Pacific Plate. One of Earth's major plates, lying west of the North American Plate and consisting of the crust and upper mantle under the Pacific Ocean.

pahoehoe. A Hawai'ian term for a type of basaltic lava typified by a relatively smooth, undulating surface of small-scale, bulbous, ropy, corded, drapery-like features.

paleontology. Study of ancient life, largely by means of fossils.

pebble. Rounded small stone, usually water-worn in the course of transport, between 0.17 and 2.5 inches in diameter.

percolation. Slow laminar movement of a fluid through small openings in a porous material.

plate. In a planetary sense, one of eight large, drifting plates composing the Earth's solid outer part, the lithosphere. Continental plates are roughly 60 miles thick.

plate tectonics. Movement and deformation caused by the interaction of large, drifting, planetary plates.

plug. Relatively small, vertical, pipe-like, cylindrical intrusive igneous body filling a volcanic vent; synonym: **volcanic neck**.

pluvial. Cooler, moister conditions in arid or semiarid areas, possibly coincident with glacial conditions in other regions.

postglacial. The interval following total disappearance of continental glaciers from nonpolar regions.

potassium-argon dating. Method of age determination in years of a rock based on radioactive decay of potassium-40 to argon-40.

pyroclastic. Said of clastic rock material formed by a volcanic explosion from a volcanic vent.

quartz. Common rock-forming mineral that is hard, chemically resistant, and composed of silicon and oxygen (SiO_2).

quartzite. A metamorphic rock formed by the recrystallization of quartz-rich sandstone.

reverse fault. A steeply inclined fault, with relative upward movement of the overlying side of the inclined fault.

rhyolite. Extrusive igneous rock (lava) of granitic composition, commonly light-colored or reddish. Relatively rich in silicon and poor in iron, magnesium, and calcium.

right-lateral fault. A fault with sideways displacement and the opposite side moving to the right. Also: right-lateral strike-slip fault, right-slip fault.

rill. A small, shoestring-like, usually ephemeral channel eroded in soft material on steep slopes by small streamlets of water, in parallel sets.

rock. Any consolidated aggregation of one or more minerals or natural glasses.

sag pond. A small body of water occupying an enclosed depression or sag formed where active or recent fault displacement has impounded drainage.

saltation. Transport of particles by wind or water in a hopping mode.

sandstone. Sedimentary rock composed primarily of rounded or angular, sand-sized particles of rock or mineral, 0.0025 to 0.08 inch in diameter.

sapping. A natural erosional process commonly involving groundwater seepage, which undercuts the base of a steep face or slope.

scarp. Topographically, a linear steep face of cliffs from a few to thousands of feet high, usually produced by faulting; abbreviated term for **escarpment.**

schist. A strongly foliated, crystalline metamorphic rock that can be readily split into thin flakes or slabs due to the well-developed parallelism of more than 50 percent of its minerals.

sediment. Solid, unconsolidated particulate matter, especially rock detritus that originates by weathering, and is transported and deposited by some agency of transport, for example, by wind or water.

sedimentary rock. Consolidated and usually cemented sediment, usually characterized by layering.

sedimentation. The process of deriving, transporting, and depositing sediment.

seismic. Pertaining to earthquakes.

shutter ridge. A landform, usually along strike-slip faults, where relatively high ground has slid alongside relatively low ground, thus shutting or damming the low area.

silica. Silicon dioxide (SiO_2).

siliceous. Said of a rock rich in silica.

silt. Fine particulate rock and mineral matter, dust-sized (finer than sand, coarser than clay), between 0.00016 and 0.0025 inch in diameter.

stalactite. Dripstone deposit built downward by water dripping from cave dwelling.

stalagmite. Dripstone deposit built upward from cave floor by water dripping from ceiling.

stratification. Layering in sedimentary rocks.

stratum. Tabular layer or bed in a sedimentary sequence of rocks (plural: **strata**).

stream capture. Natural diversion of the headwaters of one stream into a neighboring, lower level stream channel by erosion.

strike. Compass bearing of a horizontal line on the face of an inclined planar surface such as a bedding plane or fault.

subduction zone. A long narrow plate boundary where one lithospheric plate descends beneath another.

syncline. A concave-upward fold in layered rocks with younger beds toward the core, limbs inclined inward.

talus. Accumulation of angular rock fragments at the base of a steep rock face.

tectonic. Pertaining to deformation and the forces of that deformation in the Earth's crust.

tephra. Fragmental products of lava formed and ejected during volcanic eruptions.

terrace. Step-like landform consisting of a flat tread and a steep riser, commonly of river, lake, or marine origin.

terrain. Tract or region of the Earth's surface with characteristic physical features.

terrane. Large region of Earth's crust underlain by rocks of similar geologic character and history.

thrust fault. A fault with a dip of 45 degrees or less on which the upper, older block has moved upward relative to the lower, younger block.

transverse dune. A strongly asymmetric sand dune elongated perpendicularly to the prevailing wind direction.

Transverse Ranges. Physiographic province of southern California with east-west structural grain, thus transverse to the normal north-northwest trend.

travertine. A dense accumulation of calcium carbonate resulting from deposition by groundwater or surface water.

tuff. Volcanic rock formed of consolidated tephra.

unconformity. A surface of erosion or nondeposition separating younger deposits from older rock. See **angular unconformity**.

uranium. Radioactive chemical element that spontaneously decays to other elements, some stable, some radioactive.

vein. Sheet-like deposit of mineral matter within a fracture in rock.

vesicle. Small cavity of irregular to spherical shape formed by a trapped gas bubble in lava; adjective: **vesicular**.

volcanic ash. Unconsolidated, explosively fragmented, pyroclastic volcanic material of particle diameter less than 0.25 inch.

volcanic cinders. Glassy, porous fragments of lava explosively ejected from a volcanic vent, from pea to baseball size.

volcanic neck. See **plug**.

water gap. Narrow gorge or pass across resistant rock in a mountain ridge through which a stream flows.

water table. Top of the subsurface zone that is saturated with water.

wave-cut platform. Gently seaward-sloping, intertidal surface cut along a shore by long continued wave erosion.

weathering. Chemical decomposition and mechanical disintegration of rocks and minerals through the interaction with the atmosphere and biosphere.

wind gap. Saddle or notch in a ridge, a former water gap, now abandoned by the stream that cut it.

wind ripples. Wave-like, asymmetric undulations produced in wind-deposited sand.

References Cited and Further Reading

Many field guides on southern California geology have been published by a host of professional organizations, especially the South Coast Geological Society and the San Diego Association of Geologists. These guides are hard to come by because of limited editions and distribution. Some university libraries have established at least fractional collections. Local geological societies publish road guides and occasional compendia of articles about specific areas.

Atwater, T., and J. M. Stock. 1998. Pacific–North America plate tectonics of the Neogene southwestern United States: An update. *International Geology Review* 40: 373–402.

Carter, J. N., Luyendyk, B. P., and R. R. Terres. 1987. Neogene clockwise tectonic rotation of the eastern Transverse Ranges, California, suggested by paleomagnetic vectors. *Geological Society of America Bulletin* 98: 199–206.

Clifton, H. E., and R. V. Ingersoll, eds. 2010. *Geologic Excursions in California and Nevada: Tectonics, Stratigraphy, and Hydrology.* SEPM (Society for Sedimentary Geology). Pacific Section of the American Association of Petroleum Geologists.

Crouch, J. K., and J. Suppe. 1993. Late Cenozoic tectonic evolution of the Los Angeles basin and inner California borderland: A model for core complex-like extension. *Geological Society of America Bulletin* 105: 1415–34.

Ellsworth, W. L. 1990. Earthquake history, 1769–1989, chapter 6. In *The San Andreas Fault System, California*, US Geological Survey Professional Paper 1515, ed. R. E. Wallace.

Fife, D. L., and A. R. Brown, eds. 1980. *Geology and Mineral Wealth of the California Desert, Dibblee Volume.* Santa Ana, Calif.: South Coast Geological Society Guidebook.

Fife, D. L., and J. A. Minch, eds. 1982. *Geology and Mineral Wealth of the California Transverse Ranges, Mason Hill Volume.* Santa Ana, Calif.: South Coast Geological Society Guidebook.

Fritsche, A. E., Weigand, P. W., Colburn, I. P., and R. L. Harma. 2001. Transverse/Peninsular Ranges connections: Evidence for the incredible Miocene rotation. In *Geologic Excursions in Southwestern California,* Society for Sedimentary Geology, Pacific Section 89, Compilers, G. Dunne and J. Cooper, p. 101–46.

Henry, C. D., J. E. Faulds, and C. M. dePolo. 2007. Geometry and timing of strike-slip and normal faults in the northern Walker Lane, northwestern Nevada and northeastern California: Strain partitioning or sequential extensional and strike-slip deformation? In *Exhumation Associated with Continental*

Strike-Slip Fault Systems, p. 59–79. Geological Society of America Special Paper 434, eds. A. B. Till, S. M. Roeske, J. C. Sample, et al.

Hornafius, J. S. 1985. Neogene tectonic rotation of the Santa Ynez Range, western Transverse Ranges, California, suggested by paleomagnetic investigation of the Monterey Formation. *Journal of Geophysical Research: Solid Earth* 90: 12503–22.

Jahns, R. H., ed. 1954. *Geology of Southern California*. California Division of Mines and Geology Bulletin 170.

Lung, R., and R. J. Proctor, eds. 1966. *Engineering Geology in Southern California*. Association of Engineering Geologists Special Publication.

Moran, D. E., Slosson, J. E., Stone, R. O., and C. A. Yelverton, eds. 1973. *Geology, Seismicity, and Environmental Impact*. Los Angeles: University Publishers.

Nicholson, C., Sorlien, C. C., Atwater, T., Crowell, J. C., and B. P. Luyendyk. 1994. Microplate capture, rotation of the western Transverse Ranges, and initiation of the San Andreas transform as a low-angle fault system. *Geology* 22 (6): 491–95.

Norris, R. M., and R. W. Webb. 1990. *Geology of California*, Second Edition. New York: John Wiley & Sons.

Pipkin, B. W., and R. J. *Proctor*, eds. 1992. *Engineering Geology Practice in Southern California*. Association of Engineering *Geologists* Special Publication 4.

Pittman, R. 1995. *Roadside History of California*. Missoula, Mont.: Mountain Press Publishing Company.

Sylvester, A. G., and E. O. Gans. 2016. *Roadside Geology of Southern California*. Missoula, Mont.: Mountain Press Publishing Company.

Walker, J. D., Geissman, J. W., Bowring, S. A., and L. E. Babcock, compilers. 2018. *Geologic Time Scale* v. 5.0. Geological Society of America, online document, accessed 30 April 2019.

VIGNETTE 1. CRISTIANITOS FAULT

Ellis, A. J., and C. H. Lee. 1919. *Geology and Groundwaters of the Western Part of San Diego County, California*. US Geological Survey Water-Supply Paper 446.

Hart, E. W., and W. A. Bryant. 1997. *Fault-rupture Hazards Zones in California*. California Division of Mines and Geology Special Publication 42.

Kuhn, G. G., and F. P. Shepard. 1984. San Onofre and Camp Pendleton area. In *Sea Cliffs, Beaches, and Coastal Valleys of San Diego County: Some Amazing Histories and Some Horrifying Implications*. Berkeley: University of California Press, pp. 41–49.

Meehan, R. L. 1984. *The Atom and the Fault: Experts, Earthquakes, and Nuclear Power*. Cambridge, Mass.: MIT Press.

Shlemon, R. J. 1987. The Cristianitos fault and Quaternary geology, San Onofre State Beach, California. In *Centennial Field Guide Volume I*, Cordilleran Section, Geological Society of America, ed. M. L Hill, p. 171–74.

VIGNETTE 2. TORREY PINES

Abbott, P. L. 1999. *The Rise and Fall of San Diego.* San Diego: Sunbelt Publications.

Abbott, P. L., and J. A. May. 1991. *Eocene Geologic History, San Diego Region.* Los Angeles: Pacific Section of the Society of Economic Paleontologists and Mineralogists.

Grine, D. 2014. *Geology of Torrey Pines State Reserve.* Online document, accessed 11 February 2014 from http://www.torreypine.org/geology/geology.html.

Hanna, M. A. 1926. Geology of the La Jolla [15'] quadrangle, California. *University of California Publications, Bulletin of the Department of Geological Sciences* 16 (7): 187–246.

Kennedy, M. P. 1975. Geology of the San Diego metropolitan area, California, section A, western San Diego metropolitan area. *California Division of Mines and Geology Bulletin* 200: 1–39.

Ku, T. L., and J. P. Kern. 1974. Uranium-series age of the upper Pleistocene Nestor Terrace, San Diego, California. *Geological Society of America Bulletin* 85 (11): 1713–16.

Kuhn, G. G., and F. P. Shepard. 1984. *Sea Cliffs, Beaches, and Coastal Valleys of San Diego County: Some Amazing Histories and Some Horrifying Implications.* Berkeley: University of California Press.

Meldahl, K. H. 2015. *Surf, Sand, and Stone.* Berkeley: University of California Press.

VIGNETTE 3. ALISO BEACH COUNTY PARK

Ernst, W. G. 2001. Subduction, ultrahigh-pressure metamorphism, and regurgitation of buoyant crustal slices: Implications for arcs and continental growth. *Physics of the Earth and Planetary Interiors* 127: 253–75.

Fischer, P. J., and G. I. Mills. 1991. The offshore Newport–Inglewood–Rose Canyon fault zone, California: Structure, segmentation and tectonics. In *Environmental Perils, San Diego Region.* San Diego Association of Geologists, eds. P. J. Abbott and W. J. Elliott, p. 17–36.

Kamerling, M. J., and B. P. Luyendyk. 1985. Paleomagnetism and Neogene tectonics of the northern Channel Islands, California. *Journal of Geophysical Research: Solid Earth* 90 (B14): 12485–502.

Nicholson, C., Sorlien, C. C., Atwater, T., Crowell, J. C., and B. P. Luyendyk. 1994. Microplate capture, rotation of the western Transverse Ranges, and initiation of the San Andreas transform as a low-angle fault system. *Geology* 22 (6): 491–95.

Stuart, C. J. 1979. Lithofacies and origin of the San Onofre Breccia, coastal southern California. In *A Guidebook to Miocene Lithofacies and Depositional Environments, Coastal Southern California and Northwestern Baja California.* Pacific Section of the Society of Economic Paleontologists and Mineralogists, ed. C. J. Stuart, p. 25–42.

Stuart, C. J. 1979. Middle Miocene paleogeography of coastal southern California and the California borderland: Evidence from schist-bearing sedimentary rocks. In *Cenozoic Paleogeography of the Western United States*. Society of Economic Paleontologists and Mineralogists, Pacific Coast Paleogeography Symposium, Los Angeles, eds. J. M. Armentrout, M. R. Cole, and H. TerBest Jr., p. 29–44.

Woodford, A. O. 1925. The San Onofre Breccia: Its nature and origin. *California University, Department of Geological Sciences Bulletin* 15: 159–280.

VIGNETTE 4. VENTURA AVENUE ANTICLINE

Lamb, S. 2000. *Channel Islands National Park*. Western National Parks Association.

Matthews, N. E., Vazquez, J. A., and A. T. Calvert. 2015. Age of the Lava Creek supereruption and magma chamber assembly at Yellowstone based on $^{40}Ar/^{39}Ar$ and U-Pb dating of sanidine and zircon crystals. *Geochemistry, Geophysics, Geosystems* 16 (8): 2508–28.

Norris, R. M. 2003. *The Geology and Landscape of Santa Barbara County, California, and Its Offshore Islands*. Santa Barbara Museum of Natural History Monograph Number 3.

Rockwell, T. K., Keller, E. A., and G. R. Dembroff. 1988. Quaternary rate of folding of the Ventura Avenue anticline, western Transverse Ranges, southern California. *Geological Society of America Bulletin* 100 (6): 850–58.

Sylvester, A. G. 2000. Aseismic growth of Ventura Avenue anticline, southern California, 1978–1997: Evidence from precise leveling. *Surveying and Land Information Science* 60 (2): 95–108.

Yeats, R. S., and F. B. Grigsby. 1987. Ventura Avenue anticline: Amphitheater locality, California. In *Centennial Field Guide Volume I*, Cordilleran Section, Geological Society of America, ed. M. L. Hill, p. 219–23.

Zeeden, C., Rivera, T. A., and M. Storey. 2014. An astronomical age for the Bishop Tuff and concordance with radioisotopic dates. *Geophysical Research Letters* 41: 3478–84.

VIGNETTE 5. SANTA BARBARA HARBOR

Keller, E. A., and V. R. Keller. 2011. The story of Santa Barbara Harbor. In *Santa Barbara: Land of Dynamic Beauty, A Natural History*. Santa Barbara Museum of Natural History, eds. E. A. Keller and V. R. Keller.

Norris, R. M., and K. Patsch. 2005. The coast from Point Conception to Rincon Point. In *Living with the Changing California Coast*, University of California Press, eds. G. Griggs, K. Patsch, and L. Savoy, p. 373–77.

Revell, D. L., Barnard, P. L., N. Mustain, and C. D. Storlazzi. 2008. Influence of harbor construction on downcoast morphological evolution: Santa Barbara, California. In *Proceedings of the Solutions to Coastal Disasters Congress*, Turtle Bay, Oahu, Hawaii, p. 630–42.

Runyon, K., and G. Griggs. 2002. Implications of harbor dredging for the Santa Barbara littoral cell. In *California and the World Ocean*, Proceedings of American Society of Civil Engineers Conference, 2002, Santa Barbara, p. 121–35.

VIGNETTE 6. GUADALUPE-NIPOMO DUNES

Cooper, W. S. 1967. *Coastal Dunes of California*. Geological Society of America Memoir 104.

Orme, A. R. 2005. The coast from Morro Bay to Point Conception. In *Living with the Changing California Coast*, University of California Press, eds. G. Griggs, K. Patsch, and L. Savoy, p. 334–58.

Peltier, W. R., and R. G. Fairbanks. 2006. Global glacial ice volume and Last Glacial Maximum duration from an extended Barbados sea level record. *Quaternary Science Reviews* 25: 3322–37.

VIGNETTE 7. MORRO ROCK

Ernst, W. G., and C. A. Hall, Jr. 1974. Geology and petrology of the Cambria Felsite, a new Oligocene Formation, west-central California Coast Ranges. *Geological Society of America Bulletin* 85 (4): 523–32.

Fairbanks, H. W. 1904. Description of the San Luis Obispo Quadrangle, California. *Geologic Atlas. San Luis Folio 101*, US Geological Survey.

Orme, A. R. 2005. The coast from Morro Bay to Point Conception. In *Living with the Changing California Coast*, University of California Press, eds. G. Griggs, K. Patsch, and L. Savoy, p. 334–58.

Sharp, R. P., and A. F. Glazner. 1997. *Geology Underfoot in Death Valley and Owens Valley.* Missoula, Mont.: Mountain Press Publishing Company.

Wiegers, M. 2009. *Geologic Map of the Morro Bay South 7.5' Quadrangle, San Luis Obispo County, California: A Digital Database*. Sacramento: California Geological Survey. Scale 1:24,000.

VIGNETTE 8. MOONSTONE BEACH

Bailey, E. H., Irwin, W. P., and D. L. Jones. 1964. *Franciscan and Related Rocks, and Their Significance in the Geology of Western California*. California Division of Mines and Geology Bulletin 183.

Murdoch, J., and R. W. Webb. 1966. *Minerals of California: Centennial Volume.* California Division of Mines Bulletin 189.

Sinkankas, J. 1970. *Prospecting for Gemstones and Minerals*. New York: Van Nostrand Reinhold Books.

VIGNETTE 9. THE WHITTIER NARROWS

Hauksson, E., and L. M. Jones. 1989. The 1987 Whittier Narrows earthquake sequence in Los Angeles, southern California: Seismological and tectonic analysis. *Journal of Geophysical Research* 94: 9569–89.

VIGNETTE 10. WHEELER RIDGE

Bawden, G. 2001. Source parameters for the 1952 Kern County earthquake, California: A joint inversion of leveling and triangulation observations. *Journal of Geophysical Research* 106 (B1): 771–85.

Keller, E. A., Seaver, D. B., Laduzinsky, D. L., Johnson, D. L., and T. L. Ku. 2000. Tectonic geomorphology of active folding over buried reverse faults: San Emigdio Mountain front, southern San Joaquin Valley, California. *Geological Society of America Bulletin* 112 (1): 86–97.

Keller, E. A., Zepeda, R. L., Rockwell, T. K., Ku, T. L., and W. S. Dinklage. 1998. Active tectonics at Wheeler Ridge, southern San Joaquin Valley, California. *Geological Society of America Bulletin* 110 (3): 298–310.

Stein, R. S., and W. Thatcher. 1981. Seismic and aseismic deformation associated with the 1952 Kern County, California, earthquake and relationship to the Quaternary history of the White Wolf fault. *Journal of Geophysical Research* 86: 4913–28.

VIGNETTE 11. CAJON PASS

Foster, J. H. 1982. Late Cenozoic structural evolution of Cajon Valley, southern California. In *Late Cenozoic Stratigraphy and Structure of the San Bernardino Mountains,* Cordilleran Section, Geological Society of America, Guidebook 6, eds. P. M. Sadler and M. A. Kooser, p. 73–76.

Meisling, K. E., and R. J. Weldon, II. 1989. Late Cenozoic tectonics of the northwestern San Bernardino Mountains, southern California. *Geological Society of America Bulletin* 101: 106–28.

Reynolds, R. E., and R. J. Reynolds. 1994. Victorville fan and an occurrence of Sigmodon. In *Mojave Desert, Martin Stout Volume.* San Bernardino County Museum Association Annual Field Trip Guidebook 22, eds. D. Murbach and J. Baldwin, p. 153–55.

Weldon, R. J., II. 1985. Implications of the age and distribution of the late Cenozoic stratigraphy in Cajon Pass, southern California. In *Geologic Investigations along Interstate 15, Cajon Pass to Manix Lake*, San Bernardino County Museum Association, ed. R. E. Reynolds, p. 59–68.

Weldon, R.J., II. 1987. San Andreas fault, Cajon Pass, southern California, 1987. In *Centennial Field Guide Volume I*, Cordilleran Section, Geological Society of America, ed. M. L. Hill, p. 193–98.

Weldon, R. J., II, and K. E. Sieh. 1985. Holocene rate of slip and tentative recurrence interval for large earthquakes on the San Andreas fault, Cajon Pass, southern California. *Geological Society of America Bulletin* 96 (6): 793–812.

Woodburne, M. O., and D. J. Golz. 1972. *Stratigraphy of the Punchbowl Formation, Cajon Valley, Southern California.* University of California Publications in Geological Sciences 92.

VIGNETTE 12. RED ROCK CANYON

Bruns, J. J., Anderson, C. J., and D. R. Jessey. 2009. Basaltic volcanism in the southern Owens Valley, California. In *Landscape Evolution at an Active Plate Margin*, Desert Studies Consortium, California State University, eds. D. R. Jessey and R. E. Reynolds, p. 98–106.

Carr, M. D., Christiansen, R. L., Poole, F. G., and J. W. Goodge. 1997. *Bedrock Geologic Map of the El Paso Mountains in the Garlock and El Paso*

Peaks 7.5' Quadrangles, Kern County, California. US Geological Survey Map I-2389, scale 1:24,000.

Loomis, D. P., and Burbank, D. W. 1988. The stratigraphic evolution of the El Paso Basin, southern California: Implications for the Miocene development of the Garlock fault and uplift of the Sierra Nevada. *Geological Society of America Bulletin* 100 (1): 12–28.

Mabey, D. R. 1960. *Gravity Survey of the Western Mojave Desert, California.* US Geological Survey Professional Paper 316-D: 51–73.

Monastero, F. C., Sabin, A. E., and J. D. Walker. 1997. Evidence for post-early Miocene initiation of movement on the Garlock fault from offset of the Cudahy Camp Formation, east-central California. *Geology* 25 (3): 247–50.

Whistler, D. P. 2005. *Field Guide to the Geology of Red Rock Canyon and the Southern El Paso Mountains, Mojave Desert, California.* Annual meeting of Western Association of Vertebrate Paleontologists, Natural History Museum of Los Angeles County. Adopted from the field guide of National Association of Geology Teachers Far Western Section (1987).

Whistler, D. P., Tedford, R. H., Takeuchi, G. T., X. Wang, Tseng, Z. J., and M. E. Perkins. 2009. Revised Miocene biostratigraphy and biochronology of the Dove Spring Formation, Mojave Desert, California. In *Papers on Geology, Vertebrate Paleontology, and Biostratigraphy in Honor of Michael O. Woodburne*, Museum of Northern Arizona Bulletin 65, ed. L. B. Albright III, p. 331–62.

VIGNETTE 13. RAINBOW BASIN

Burke, D. B., Hillhouse, J. W., McKee, E. H., Miller, S. T., and J. L. Morton. 1982. Cenozoic rocks in the Barstow Basin area of southern California: Stratigraphic relations, radiometric ages, and paleomagnetism. *US Geological Survey Bulletin* 1529-E: E1-E16.

Ingersoll, R. V., Devaney, K. A., Geslin, J. K., Cavazza, W., Diamond, D. S., Heins, W. A., Jagiello, K. J., Marsaglia, K. M., Paylor, E. D., II, and P. F. Short. 1996. *The Mud Hills, Mojave Desert, California: Structure, Stratigraphy, and Sedimentology of a Rapidly Extended Terrane.* Geological Society of America Special Paper 103.

MacFadden, B. J., Swisher, C. C., III, Opdyke, N. D., and M. O. Woodburne. 1990. Paleomagnetism, geochronology, and possible tectonic rotation of the middle Miocene Barstow Formation, Mojave Desert, southern California. *Geological Society of America Bulletin* 102: 478–93.

Woodburne, M. O., ed. 2004. *Late Cretaceous and Cenozoic Mammals of North America: Biostratigraphy and Geochronology.* New York: Columbia University Press.

Woodburne, M. O., Tetford, R. H., and C. C. Swisher III. 1990. Lithostratigraphy, biostratigraphy, and geochronology of the Barstow Formation, Mojave Desert, southern California. *Geological Society of America Bulletin* 102: 459–77.

VIGNETTE 14. AMBOY AND PISGAH CRATERS

Glazner, A. F., Farmer, G. L., Hughes, W. T., Wooden, J. L., and W. Pickthorn. 1991. Contamination of basaltic magma by mafic crust at Amboy and Pisgah Craters, Mojave Desert, California. *Journal of Geophysical Research* 96: 13673–91.

Greely, R. 1990. Amboy, California. In *Volcanoes of North America: United States and Canada*, Cambridge University Press, eds. C. A. Wood and J. Kienle, p. 243–45.

Phillips, F. M. 2003. Cosmogenic ^{36}Cl ages of Quaternary basalt flows in the Mojave Desert, California, USA. *Geomorphology* 53: 199–208.

Sweet, W. E., Jr. 1980. Geology and genesis of hectorite, Hector, California. In *Geology and Mineral Wealth of the California Transverse Ranges, Mason Hill Volume*, South Coast Geological Society Symposium and Guidebook Number 10, Santa Ana, California, eds. D. L. Fife and J. A. Minch, p. 279–83.

Sylvester, A. G., Burmeister, K. C., and W. S. Wise. 2002. Faulting and effects of associated shaking at Pisgah Crater volcano caused by the 16 October 1999 Hector Mine earthquake (M_w 7.1), central Mojave Desert, California. *Bulletin of the Seismological Society of America* 92: 1333–40.

Wise, W. S. 1966. Geologic map of the Pisgah and Sunshine cone lava fields, California. *NASA Technical Letter* 11.

Wise, W. S. 1969. Origin of basaltic magmas in the Mojave Desert area, California. *Contributions to Mineralogy and Petrology* 23 (1): 53–64.

VIGNETTE 15. LAKE MANIX

Enzel, Y., Wells, S. G., and N. Lancaster, 2003. Late Pleistocene lakes along the Mojave River, southeast California. In *Paleoenvironments and Paleohydrology of the Mojave and Southern Great Basin Deserts*, Geological Society of America Special Paper 368, eds. Y. Enzel, S. G. Wells, and N. Lancaster.

Meek, N. 1989. Geomorphic and hydrologic implications of the rapid incision of Afton Canyon, Mojave Desert, California. *Geology* 17 (1): 7–10.

Orme, A. R. 2008. Lake Thompson, Mojave Desert, California: The late Pleistocene lake system and its Holocene desiccation. *Geological Society of America Special Papers* 439: 261–78.

Owen, L. A., Finkel, R. C., Minch, R. A., and A. E. Perez. 2003. Extreme southwestern margin of late Quaternary glaciation in North America: Timing and controls. *Geology* 31 (8): 729–32.

Reheis, M. C., Miller, D. M., McGeehin, J. P., Redwine, J. R., Oviatt, C. G., and J. E. Bright. 2015. Directly dated MIS 3 lake-level record from Lake Manix, Mojave Desert, California, USA. *Quaternary Research* 83 (1): 187–203.

Reheis, M. C., and J. L. Redwine. 2008. Lake Manix shorelines and Afton Canyon terraces: Implications for incision of Afton Canyon. In *Late Cenozoic Drainage History of the Southwestern Great Basin and Lower Colorado River Region: Geologic and Biotic Perspectives*, Geological Society of America Special Paper 439, eds. M. C. Reheis, R. Hershler, and D. M. Miller, p. 227–59.

Reynolds, R. E., and D. M. Miller, eds. 2010. *Overboard in the Mojave: 20 Million Years of Lakes and Wetlands*. Fullerton: Desert Studies Consortium, California State University.

Seiple, E. 1994. Plant and animal life of Pleistocene Lake Manix. *California Geology* 47 (2): 50–57.

Sharp, R. P., Allen, C. A., and M. F. Meier. 1959. Pleistocene glaciers on southern California mountains. *American Journal of Science* 257 (2): 81–94.

VIGNETTE 16. THE BLACKHAWK LANDSLIDE

Ericksen, G. E., G. Plafker, and J. F. Concha. 1970. *Preliminary Report on the Geologic Events Associated with the May 31, 1970, Peru Earthquake*. US Geological Survey Circular 639.

Melosh, H. J. 1983. Acoustic fluidization. *American Scientist* 71: 158–65.

Shreve, R. L. 1987. Blackhawk landslide, southwestern San Bernardino County, California. In *Centennial Field Guide Volume I*, Cordilleran Section, Geological Society of America, ed. M. L. Hill, p. 109–14.

Smith, R. B., and L. J. Siegel. 2000. *Windows into the Earth*. New York: Oxford University Press.

Stout, M. L. 1982. Age and engineering geologic observations of the Blackhawk landslide, southern California. In *Geology and Mineral Wealth of the California Transverse Ranges, Mason Hill Volume*, South Coast Geological Society Symposium and Guidebook Number 10, Santa Ana, California, eds. D. L. Fife and J. A. Minch, p. 630–33.

Woodford, A. O., and T. F. Harriss. 1928. Geology of Blackhawk Canyon, California. *California University Publications in Geological Sciences Bulletin* 17 (8): 265–304.

Yarnold, J. C., and J. P. Lombard. 1989. Facies model for large rock-avalanche deposits formed in dry climates. In *Conglomerates in Basin Analysis: A Symposium Dedicated to A. O. Woodford*. Pacific Section of the Society of Economic Paleontologists and Mineralogists, Bakersfield, California, eds. I. P. Colburn, P. L. Abbott, and J. A. Minch, p. 9–31.

VIGNETTE 17. KELSO DUNES

Lancaster, N. 1993. Development of Kelso Dunes, Mojave Desert, California. *National Geographic Research and Exploration* 9: 444–59.

Lindsay, J. F., Criswell, D. R., Criswell, T. L., and B. S. Criswell. 1976. Sound producing dune and beach sands. *Geological Society of America Bulletin* 87: 463–73.

Paisley, E. C. I., Lancaster, N., Gaddis, L. R., and R. Greeley. 1991. Discrimination of active and inactive sand from remote sensing: Kelso Dunes, Mojave Desert, California. *Remote Sensing of Environment* 37: 153–66.

Ramsey, M. S., Christensen, P. R., Lancaster, N., and D. A. Howard. 1999. Identification of sand sources and transport pathways at the Kelso Dunes, California, using thermal infrared remote sensing. *Geological Society of America Bulletin* 111: 646–62.

Sharp, R. P. 1966. Kelso Dunes, Mojave Desert, California. *Geological Society of America Bulletin* 77: 1045–74.

Trexler, D. T., and W. N. Melhorn. 1986. Singing and booming sand dunes of California and Nevada. *California Geology* 39 (7): 147–52.

VIGNETTE 18. MITCHELL CAVERNS

Aalbu, R. 1989. An analysis of the Coleoptera of Mitchell Caverns, San Bernardino County, California. *National Speleological Bulletin 51 (1): 1–10.*

Burgess, S. D., Muirhead, J. D., and S. A. Bowring. 2017. Initial pulse of Siberian Traps sills as the trigger of the end-Permian mass extinction. *Nature Communications* 8 (164): 1–6.

McCurry, M. 1988. Geology and petrology of the Woods Mountains volcanic center, southeastern California: Implications for the genesis of peralkaline rhyolite ash flow tuffs. *Journal of Geophysical Research* 93 (B12): 14,835–55.

Norris, R. M. 1985. Geologic Setting of Mitchell Caverns in Mitchell Caverns Natural Preserve in the Providence Mountains State Recreation Area. *California Geology 38 (2): 34–38.*

VIGNETTE 19. THE SAN ANDREAS FAULT

Baldwin, E. J., Foster, J. H., Lewis, W. L. Gath, E., Hardy, J. K., Raub, M. L. and M. Rendina, eds. 1989. *San Andreas Fault: Cajon Pass to Wallace Creek, Volume 2.* Field Trip Guidebook No. 17, Santa Ana, Calif.: South Coast Geological Society.

Baldwin, E. J., Foster, J. H., Lewis, W. L., and J. K. Hardy, eds. 1989. *San Andreas Fault: Cajon Pass to Wallace Creek, Volume 1.* Field Trip Guidebook No. 17, Santa Ana, Calif.: South Coast Geological Society.

Ely, M. F. 1982. Erosion activity during 1978 on the Heath Canyon landslide, Wrightwood area, San Bernardino County, California. In *Geology and Mineral Wealth of the California Transverse Ranges, Mason Hill Volume,* South Coast Geological Society Symposium and Guidebook Number 10, Santa Ana, California, eds. D. L. Fife and J. A. Minch, p. 637–44.

Fairbanks, H. W. 1907. The great earthquake rift of California. *Bulletin of the California Physical Geography Club* 10: 321–37.

Hough, S. E. 2004. *Finding Fault in California.* Missoula, Mont.: Mountain Press Publishing Company.

Iacopi, R. 1971. *Earthquake Country.* Menlo Park, Calif.: Lane Books.

Jacoby, G. C., Shepard, P. R., and K. E. Sieh. 1988. Irregular recurrence of large earthquakes along the San Andreas fault: Evidence from trees. *Science* 241 (4862): 196.

Lynch, D. K. 2009. *Field Guide to the San Andreas Fault.* Topanga, Calif.: Thule Scientific.

Meisling, K. E., and K. E. Sieh. 1980. Disturbance of trees by the 1857 Fort Tejon earthquake, California. *Journal of Geophysical Research: Solid Earth* 85: 3225–38.

Scharer, K. M., Weldon, R. J., II., Fumal, T. E., and G. Biasi. 2007. Paleoearthquakes on the southern San Andreas fault, Wrightwood, California, 3000 to 1500 B.C.: A new method for evaluating paleoseismic evidence and earthquake horizons. *Bulletin of the Seismological Society of America* 97 (4): 1054–93.

Schubert, C. 1982. Neotectonics of a segment of the San Andreas fault, southern California (USA). *Eiszeitalter und Gegenwart* 32: 13–22.

Sharp, R. P., and L. H. Noble. 1953. Mudflow of 1941 at Wrightwood, southern California. *Geological Society of America Bulletin* 53 (5): 547–60.

Sieh, K. E. 1978. Prehistoric large earthquakes produced by slip on the San Andreas fault at Pallett Creek, California. *Journal of Geophysical Research: Solid Earth* 83 (B8): 3907–39.

Sylvester, A. G., and J. C. Crowell. 1989. *The San Andreas Transform Belt.* 28th International Geological Congress Field Trip Guidebook T309.

Weldon, R. J., II. 1987. San Andreas fault, Cajon Pass, southern California, 1987. In *Centennial Field Guide Volume I*, Cordilleran Section, Geological Society of America, ed. M. L. Hill, p. 193–98.

Weldon, R. J., II, Scharer, K. M., Fumal, T., and G. Biasi. 2004. Wrightwood and the earthquake cycle: What a long recurrence record tells us about how faults work. *GSA Today* 14 (9): 4–10.

VIGNETTE 20. DEVILS PUNCHBOWL

Barrows, A. G., Kahle, J. E., and D. J. Berry. 1985. Earthquake hazards and tectonic history of the San Andreas fault zone, Los Angeles County, California. *California Division of Mines and Geology Open File Report* 85-10: 111–39.

Chester, F. M. 1999. *Field Guide to the Punchbowl Fault Zone at Devils Punchbowl, Los Angeles County Park, California*. Online document, accessed 15 November 2018 at http://seismo.berkeley.edu.

Chester, F. M., and J. S. Chester. 1998. Ultracataclasite structure and friction processes of the Punchbowl fault, San Andreas system, California. *Tectonophysics* 29: 199–221.

Coffey, K. T., Ingersoll, R. V., and A. K. Schmitt. 2019. Stratigraphy, provenance, and tectonic significance of the Punchbowl block, San Gabriel Mountains, California, USA. *Geosphere* 15 (2): 479–501.

Dibblee, T. W., Jr. 1987. Geology of the Devils Punchbowl, Los Angeles County, California. In *Centennial Field Guide, Volume I*, Cordilleran Section, Geological Society of America, ed. M. L. Hill, p. 207–10.

Noble, L. F. 1953. *Geology of the Pearland Quadrangle, California*. US Geological Survey Map GQ-24, scale 1:24,000.

Noble, L. F. 1954. *Geology of the Valyermo Quadrangle and Vicinity*. US Geological Survey Map GQ-50, scale 1:24,000.

Pelka, G. J. 1971. Paleocurrents of the Punchbowl Formation and their interpretation. *Geological Society of America Abstracts with Programs* 3 (2): 176.

Woodburne, M. O. 1975. Late Tertiary nonmarine rocks, Devils Punchbowl and Cajon Valley, Southern California. *California Division of Mines and Geology Bulletin* 118: 187–96.

Woodburne, M. O., and D. J. Golz. 1972. *Stratigraphy of the Punchbowl Formation, Cajon Valley, southern California.* University of California Publications in Geological Sciences 92.

Wright, L. A., and B. W. Troxel. 2002. *Levi Noble: Geologist. His Life and Contributions to Understand the Geology of Death Valley, the Grand Canyon, and the San Andreas Fault.* USGS Open File Report 02-422.

VIGNETTE 21. ST. FRANCIS DAM

Clements, T. 1982. St. Francis Dam failure of 1928. In *Geology and Mineral Wealth of the California Transverse Ranges, Mason Hill Volume,* South Coast Geological Society Symposium and Guidebook Number 10, Santa Ana, Calif., eds. D. L. Fife and J. A. Minch, p. 657–60.

Davis, M. L. 1993. *Rivers in the Desert.* New York: HarperCollins.

Outland, C. F. 1977. *Man-Made Disaster: The Story of St. Francis Dam*, rev. ed., Glendale, Calif.: Arthur H. Clark.

Paling, D. M. 2001. *The Rise and Fall of the St. Francis Dam.* MS thesis, Denver: University of Colorado.

Rogers, J. D. 1992. Reassessment of the St. Francis Dam failure. In *Engineering Geology Practice in Southern California*, Association of Engineering Geologists, Southern California Section, Special Publication No. 4, eds. B. W. Pipkin and R. J. Proctor, p. 639–66.

Rogers, J. D. 2015. *St. Francis Dam: What Mulholland Did That Led to the Failure.* Symposium on the St. Francis Dam Failure, Santa Clarita Historical Society and Cal State Northridge Anthropology Department, online document, accessed 15 November 2018 at https://scvtv.com.

Rogers, J. D. 2017. *Who Designed the Ill-Fated St. Francis Dam?* World Environmental and Resources Council, online document, accessed 15 November 2018 at https://scvhistory.com.

St. Francis Dam Disaster, Water and Power Associates, online document, accessed 15 November 2018 at http://waterandpower.org.

Wilkman, J. 2016. *Floodpath: The Deadliest Man-Made Disaster of Twentieth-Century America and the Making of Modern Los Angeles.* New York: Bloomsbury Press.

INDEX

Page numbers in bold include photographs.

Art Sylvester (left) and Allen Glazner (right).

ARTHUR GIBBS SYLVESTER was born, raised, and educated in southern California. After earning his BA in liberal arts at Pomona College and PhD in geology at UCLA, he joined a team of Shell Development Company research geologists to study the tectonic history of the Pacific margin of the United States. UC Santa Barbara lured him from Shell to teach courses in structural geology, field geology, and petrology. He led more than 300 field trips in southern California for student, industrial, and professional geologists. After retiring from active teaching in 2003, he and Libby O'Black Gans coauthored *Roadside Geology of Southern California*.

ROBERT P. SHARP (1911–2004) co-conceived of and cowrote the first edition of this book when he was professor emeritus at the California Institute of Technology (Caltech), where he taught generations of geology students, leading them on field trips across the state. He was an expert in sand dune physics and also planetary sciences. He received the Penrose Medal from the Geological Society of America, its highest honor, in 1977.

ALLEN F. GLAZNER, Bob Sharp's coauthor on the first edition, also earned a BA at Pomona College and a PhD in geology from UCLA and is currently professor emeritus of geology at the University of North Carolina at Chapel Hill. A native Californian, he has conducted geological research in the Sierra Nevada and the Mojave Desert since his undergraduate days. He is also the coauthor of *Geology Underfoot in Death Valley and Owens Valley* and *Geology Underfoot in Yosemite National Park*.

ELIZABETH O'BLACK GANS, Libby to most, earned a BS in geology from UC Santa Barbara, where she was introduced to plenty of awesome southern California geology, as well as to the art of map making. Shortly after graduation, Libby started Gans Illustrations and began working on maps and scientific illustrations for publication, and was the first digital artist for the Dibblee Geological Foundation. Libby illustrated and co-authored *Roadside Geology of Southern California* in 2016.